ARCHITECTING THE AGENTIC ENTERPRISE

IDENTITY, AUTHORITY, GOVERNANCE, AND OPERATIONS FOR AUTONOMOUS AI SYSTEMS

T. BRENT LONG

Architecting the Agentic Enterprise

Identity, Authority, Governance, and Operations for Autonomous AI Systems

The information in this book is provided for educational and informational purposes. Technology, security threats, standards, protocols, products, and recommended practices change over time. Although reasonable care has been taken in preparing this work, neither the author nor the publisher makes any representation or warranty concerning the completeness, accuracy, suitability, or continued applicability of its contents.

Nothing in this book constitutes legal, regulatory, compliance, financial, or cybersecurity advice. Readers are responsible for evaluating the material within the context of their own systems, organizations, obligations, and risk environments. Implementation decisions should be reviewed by appropriately qualified technical, security, legal, and compliance professionals.

Architectural examples, scenarios, organizations, systems, agents, transactions, and identities described herein are illustrative unless otherwise identified. Any resemblance to real organizations, systems, or events is coincidental. Examples are not intended to represent full production implementations or guarantees of security, reliability, compliance, or fitness for a particular purpose.

Product names, company names, standards, protocols, and trademarks referenced in this book remain the property of their respective owners. Their inclusion is for identification and educational purposes and does not imply sponsorship, endorsement, or affiliation. The author and publisher disclaim liability for any loss, damage, security incident, service interruption, compliance failure, or other consequence arising directly or indirectly from the use or misuse of the information contained in this publication.

Related title by this author:

Modern Cloud Architecture in the Age of AI

Published by Pentesoft Press

Printed in the United States of America

ISBN: 979-8-952677-03-6

7 6 5 4 3 2 1

Contents

0. Introduction 5
1. The Delegated Actor 15
2. Establishing Agent Identity 29
3. Delegating Authority 43
4. Bounding Agent Authority 57
5. Tools, Data, and the Execution Boundary 75
6. Deterministic Control and Human Authority 93
7. Interruption, Revocation, and Containment 111
8. Decision Trace and Evidence 125
9. Evaluation, Operations, and Failure 141
10. Governing the Agentic Enterprise 153
11. Following the Authority 163
12. A Final Word From the Author 173
Appendix A: Glossary 177
Appendix B: References 185
Appendix C: Implementation Guides 189
Azure 191
Amazon Web Services (AWS) 194
Google Cloud Platform (GCP) 196
One sequence, any cloud 198
Appendix D: Architectural Chain Review Card 199

INTRO 0

Introduction

Software has been taking actions for a long time.

Payroll systems transfer money. Trading platforms buy and sell securities. Workflow engines approve requests and route work. Industrial control systems open valves and change set points. Scheduled processes create invoices, suspend accounts, move files, and modify databases without waiting for a person to click a button. There is nothing particularly new about software producing consequences in the physical or digital world.

What is changing is how those actions are selected.

Traditional software generally operates within a decision structure designed in advance. Even when the system is complicated, its behavior is substantially determined by code, rules, state machines, workflow definitions, configuration, or some combination of them. A developer may not be able to predict every runtime condition, but the mechanisms by which the software responds to those conditions have been deliberately constructed.

An AI agent introduces a different relationship between objective and action.

Instead of being given only a sequence of steps to execute, an agent may be given a goal and allowed to determine how to pursue it. It can interpret a request, gather information, choose among available tools, evaluate intermediate results, revise its approach, and decide what to do next. Depending on the architecture, it may continue this process until it concludes that the task has been completed.

That difference is easy to underestimate because many early uses of generative AI do not cross an important architectural boundary. A model that summarizes a document, drafts an email, answers a question, or recommends a course of action may produce an incorrect result, but the result remains information. A human or another deterministic process still decides whether that information becomes an action.

The architecture changes when the system can act on the recommendation itself.

Consider a procurement system. An AI model might analyze purchasing history and recommend that an employee order a replacement laptop. That creates familiar concerns about accuracy, data protection, and model quality. But suppose the system can also identify an approved supplier, obtain current pricing, construct a purchase order, and submit it.

The problem is no longer whether the model produced a good answer.

The system has exercised authority.

Later chapters use one request. Jesse Fox, a purchaser in the Portland office, asks the procurement agent to replace a failed laptop. The approved catalog item costs $3,850 from an approved vendor, billed to cost center CC-4410. Departmental policy lets the agent complete an approved-category purchase at or below $4,000. Above that it must prepare the order and wait. When a later chapter changes a number, it is this order changing.

The same distinction appears almost everywhere agents are being introduced. A coding assistant can suggest a source-code change, or an agent can modify a repository and create a pull request. It can recommend a deployment, or it can initiate one. A customer-service model can draft a response explaining that a customer deserves a refund, or an agent can issue the refund. A financial agent can identify an anomalous transaction, or it can freeze an account. An administrative agent can explain how to change a user's access, or it can invoke the identity system and change it.

In each pair, the intelligence may be similar. The architectural consequence is not.

The important transition is therefore not from software without AI to software with AI. It is the transition from **inference to consequence.**

A useful way to think about that transition is as a progression. A system may observe information. It may infer something from that information. It may recommend an action. It may decide that an action should occur. Finally, it may possess sufficient access and authority to cause that action to occur.

Each step changes the risk, but the final step changes the architecture.

Once an AI-enabled system can exercise consequential authority, questions that might previously have been treated independently become tightly connected. Who requested the work? Which software is actually executing it? Which agent is responsible for the decision? What authority was granted to that agent? Did the authority come from the requesting user, from an organizational role, or from a standing service credential? Which resources may the agent access? Which actions may it perform? What prevents it from doing something outside the intended task? What happens if it delegates part of the work to another agent? What happens if the agent needs to be stopped? And if an unexpected transaction is discovered tomorrow, can the enterprise reconstruct why it occurred?

These are not primarily model questions.

They are architecture questions.

The Agent Is Not the Model

One source of confusion is the tendency to use *model* and *agent* almost interchangeably.

A model can contribute reasoning, classification, generation, planning, or other inference capabilities to an agent, but the model is not the complete actor. The agent exists in a larger execution environment that provides instructions, context, memory, tools, credentials, policies, orchestration, and connections to enterprise systems. The model may help decide what should happen next, but other parts of the architecture determine what the system is capable of doing.

Security and governance problem does not stop at the model boundary. An enterprise may carefully evaluate a model and still build an unsafe agent around it. Conversely, an imperfect model can sometimes operate acceptably when the architecture sharply limits what its errors can affect. The model may propose an inappropriate action, but an independent policy control can prevent the action. A transaction limit can constrain its consequence. An approval requirement can introduce additional authority. A short-lived credential can expire. A rate limit can prevent a mistake from becoming a thousand mistakes.

The architecture therefore cannot be based on an assumption that sufficient model quality will eventually eliminate the need for these controls.

Probabilistic reasoning and deterministic control serve different purposes.

This becomes particularly important when agents are described as being *aligned*, *trusted*, or *instructed* not to perform certain actions. Instructions matter. They can substantially influence behavior. But instructions to a model are not equivalent to authorization policy enforced by the enterprise systems upon which the agent acts.

A prompt is not a security boundary.

A tool description is not an authorization policy.

Model confidence is not permission.

If an action is prohibited, the architecture should not depend solely upon the agent deciding not to attempt it.

Identity Does Not Answer the Hardest Question

Enterprise architects already have mechanisms for identifying software. Applications use service accounts. Cloud workloads obtain workload identities. Services authenticate to other services. Certificates, tokens, secrets, and platform identity systems allow one component to establish the identity of another.

Those mechanisms remain important in agentic systems. They do not, by themselves, solve the agent authority problem.

Suppose an enterprise can establish with complete confidence that a particular request came from Agent A running in Workload B. That answers an identity question: *What is acting?*

It does not necessarily answer the authority question: *Why is it allowed to do this?*

The distinction becomes particularly important when the agent is acting because a human asked it to perform a task. The human has an identity. The runtime has an identity. The logical agent may have an identity. A particular executing instance of that agent may also need to be distinguished from every other instance.

None of those identities should automatically be treated as interchangeable.

The user is not the agent. The agent is not the model. The agent is not necessarily the runtime. And authenticating any of them does not establish all of the authority required for a particular action.

This is where delegation becomes central.

A human employee may have authority to approve purchases up to a particular amount. That does not necessarily mean an agent working for the employee should receive every permission the employee possesses. The employee may have access to salary information, customer records, administrative systems, or transactions that have nothing to do with the task being delegated.

Giving the agent the user's complete standing authority would be convenient.

It would also be difficult to reconcile with the tenet of least privilege.

The alternative is to treat agent authority as something deliberately delegated and bounded: authority appropriate to a particular task, resource, action, context, duration, and level of consequence. The agent can then act on behalf of a legitimate principal without becoming indistinguishable from that principal.

This is more than a credential-management problem. It is an architectural relationship that must survive throughout execution.

The Problem of Delegated Agency

This book begins from a simple thesis:

An AI agent is a delegated actor whose identity, authority, context, tools, decisions, actions, and accountability must remain connected throughout its execution.

The word *delegated* is important.

Enterprises ultimately decide what their systems are permitted to do. Even highly autonomous agents operate because some person or organization has placed them into an environment, supplied them with capabilities, and authorized some range of activity. Autonomy changes how frequently humans participate in individual decisions. It does not eliminate the organizational source of authority.

The word *actor* is equally important.

An agent is architecturally interesting because it can do more than calculate a result. Within its assigned capabilities, it can select actions and initiate them. That makes the agent part of the enterprise's authorization and accountability architecture.

This leads to the governing question for the remainder of this book:

Who authorized this agent to do what, using which authority, within what boundaries, and how can the enterprise prove afterward what occurred?

A mature agentic architecture should be able to answer that question for consequential actions.

Not merely which API key was used.

Not merely which service account authenticated.

Not merely which model generated the request.

Not merely which user originally opened the application.

The architecture needs to preserve the relationship among all of them.

Preserving the Chain

The architectural model developed throughout this book can be represented as a chain:

Principal → Delegation → Agent Identity → Task → Policy → Tool → Action → Outcome → Evidence

The **principal** is the legitimate source of authority: perhaps a person, an organization, or another authorized actor.

Delegation establishes what authority is being entrusted for the work and under which constraints.

Agent identity establishes the actor to which that authority has been entrusted and, where necessary, the particular executing instance.

The **task** establishes the purpose and scope within which that authority is being exercised.

Policy determines whether a proposed use of authority is permitted under the applicable enterprise rules and context.

A **tool** provides the capability through which the agent can access information or affect enterprise state.

An **action** is the consequential operation actually attempted or performed.

The **outcome** is what happened as a result.

And **evidence** preserves enough information to connect the outcome back through the action, tool, policy decision, task, agent, delegation, and originating principal.

This is not a prescription for nine products, services, database tables, tokens, or network hops. Some implementations will combine several of these responsibilities. Others will distribute them across platforms and organizational boundaries. Technologies and protocols will change.

The architectural requirement is continuity.

If the principal becomes disconnected from the delegation, the enterprise may know what acted without knowing whose authority it exercised. If delegation becomes disconnected from the task, a narrow request can become broad standing authority. If policy exists only inside the agent's reasoning process, the actor being constrained becomes responsible for enforcing its own limits. If an action becomes disconnected from its outcome, the enterprise cannot reliably determine whether execution succeeded. If evidence records the API call but not the authority chain that produced it, an audit trail may show what happened without establishing why it was permitted to happen.

Each broken connection creates a different class of architectural weakness.

Agentic systems make these weaknesses more consequential because agents can combine capabilities dynamically. A conventional application may expose a finite set of workflows designed in advance. An agent may possess several individually reasonable capabilities—searching customer information, creating documents, sending messages, submitting transactions—and decide at runtime how to combine them.

That flexibility is part of the value of agency.

It is also why authority must be bounded outside the agent's reasoning.

The chain will be built one link at a time in the chapters that follow. Appendix D returns to it as a review card: one figure, a forward walk, a reverse walk, and a way to treat a missing connection as a finding rather than as a documentation gap. Use that appendix when the architecture is on the table. Use the chapters when you need to know why a particular link is there.

Autonomy Is Production Privilege

Much of the current discussion about agents treats autonomy as a capability milestone. Systems progress from answering questions, to using tools, to completing tasks, to coordinating other agents. Greater autonomy is often described as though it were simply greater technical sophistication.

For an enterprise architect, there is another way to view the same progression.

Increasing autonomy is increasing production privilege.

An agent that can only read information has one level of consequence. An agent that can recommend a change has another. An agent that can prepare a transaction for approval has more. An agent that can execute transactions within defined limits has more still. An agent that can independently delegate work to other actors and initiate high-consequence actions possesses substantial operational authority.

The appropriate question is therefore not, *How autonomous can we make the agent?*

It is, *How much authority can this architecture safely support?*

That answer depends on more than model performance. It depends on identity, delegation, policy enforcement, transaction boundaries, observability, interruption, evidence, operational maturity, and governance.

An enterprise should not grant autonomy that exceeds its ability to constrain, observe, or interrupt. Further, it should not grant autonomy whose consequences it cannot reconstruct afterward.

This does not imply that humans must approve every consequential action. Such a requirement would eliminate much of the practical value of autonomous systems and would frequently produce approval processes in which humans mechanically accept recommendations they have neither the time nor information to evaluate.

Human involvement is one control among many.

The deeper requirement is that authority remain bounded by consequence.

Low-consequence, reversible actions may justify considerable autonomy. Higher-consequence actions may require tighter transaction limits, stronger policy, additional evidence, separation of duties, or explicit approval. Some actions may remain inappropriate for autonomous execution regardless of model quality.

The objective is not maximum autonomy. It is governable autonomy.

CHAPTER 1

The Delegated Actor

A purchase order has just been created.

Jesse Fox, a purchaser in the Portland office, asked for a replacement laptop. An AI agent selected the approved catalog item, checked inventory, found none on hand, confirmed the approved vendor and the $3,850 price, verified that the order was within the $4,000 limit for cost center CC-4410, and submitted the purchase order.

Who created the purchase order?

The employee initiated the work, but did not create it. The language model helped determine what to do, but a model is not normally the component that authenticates to a procurement system and submits a transaction. The agent runtime executed software, but the runtime may host many agents performing unrelated work. A procurement API accepted the request, but it did not decide that a laptop should be purchased.

We could simply say that "the AI" created the order, but that phrase collapses several architecturally different things into one convenient noun.

That becomes a problem as soon as the enterprise asks the questions that follow.

Which identity appeared in the procurement logs? Whose authority permitted the purchase? Was that authority associated with the employee, the agent, the runtime, or a service account? Which particular execution made the decision? What was it authorized to purchase? Could another agent running on the same platform have exercised the same authority? If the transaction was incorrect, which component should have prevented it? If an investigator returned six months later, could the organization reconstruct what happened?

Before answering those questions, we need to identify the actor.

Finding the Actor

The current discussion around AI agents tends to begin with capabilities: reasoning, planning, memory, tool use, orchestration, retrieval, multi-agent coordination, and increasingly sophisticated models. Those capabilities matter, but they do not provide a sufficiently precise enterprise architecture.

The National Institute of Standards and Technology has begun examining the problem from a different direction. In February 2026, NIST's National Cybersecurity Center of Excellence published the initial public draft of *Accelerating the Adoption of Software and Artificial Intelligence Agent Identity and Authorization*. The work focuses specifically on applying identity standards and established security practices to software agents, particularly agentic AI applications.

Its status is important. This is a concept paper and potential NCCoE project, not a finalized NIST standard or practice guide. But the problem NIST is investigating is directly relevant to enterprise architecture: organizations need ways to identify software agents and control the authority they exercise as those agents gain access to data, tools, applications, and other resources.

NIST describes software and AI agents in this context as systems capable of autonomous decision-making and action with limited human supervision in pursuit of complex goals. Its work explicitly considers identification, authentication, authorization, delegation, auditing, non-repudiation, logging, and transparency.

Those concerns immediately move the discussion beyond the model.

If an enterprise must authenticate something, authorize its access, trace its delegation, and audit its actions, then it needs to know what that something is.

That sounds obvious. In practice, it is not.

An operational agentic system may contain a human principal, an application, an agent runtime, a logical agent, one or more executing agent instances, one or more models, memory and retrieval services, tools, APIs, credentials, policy controls, and downstream enterprise systems. Depending on the implementation, several of those may share infrastructure or identities.

Calling the collection "the agent" may be convenient when discussing functionality.

It is inadequate when discussing authority.

The Model

Begin with the easiest distinction.

The model is not the agent.

A model provides capabilities that an agent may use. It can interpret instructions, classify information, extract facts, generate content, evaluate alternatives, formulate plans, select among possible actions, or reason about the result of a previous action.

Those capabilities may make the agent useful. They do not by themselves constitute the complete actor.

Imagine that the procurement agent uses a model to determine that the employee needs a particular class of laptop. Six months later, the enterprise replaces that model with another one. The procurement agent still exists. Its business purpose has not necessarily changed. Its permitted tools may remain the same. Its organizational owner may remain the same. Its policies and transaction limits may remain the same.

But the reasoning component changed.

The actor did not necessarily become a different actor.

This is similar to distinctions enterprise architects already make elsewhere. A database is not the application that uses it. A rules engine is not the business process that invokes it. A cryptographic library is not the service whose requests it protects.

The model is a powerful dependency of the agent, but the architectural identity of the agent should not be casually reduced to the identity of its model.

This becomes even clearer when one agent uses multiple models. It may route different tasks to different models based on capability, latency, sensitivity, cost, or availability. If changing the model automatically changes the identity of the actor, the architecture quickly becomes incoherent.

The model helps answer the question: *How did the system decide?*

It does not, by itself, answer: *Who acted?*

The Runtime

The runtime presents a subtler problem because enterprise infrastructure already has well-developed ways to identify executing workloads.

Suppose the procurement agent runs inside a container in a managed platform. The container can be identified. The workload may receive a cryptographically verifiable identity. Infrastructure policy may determine which services it can reach. Credentials can be issued to it. Its network traffic and telemetry can be associated with that workload.

All of that is valuable.

But consider what happens if the runtime hosts several agents.

The procurement agent can submit purchase requests. A customer-service agent can retrieve customer records. A software-development agent can interact with source repositories. All three may execute within the same agent platform or orchestration environment.

The platform's identity answers an important question: *What workload is executing this software?*

It does not necessarily answer: *Which delegated actor is performing this task?*

Those questions may occasionally have the same answer. Architecturally, we should not assume that they do.

The distinction becomes more important as agent platforms become shared enterprise infrastructure. A runtime may start and stop instances, schedule work, manage model access, maintain sessions, provide common tool connectivity, and support many agents. Its security identity describes the executing environment. The agent's identity describes the logical actor operating within it.

Confusing the two can lead to a dangerous shortcut: granting the runtime all of the authority required by every agent it might execute.

That makes infrastructure convenient.

It also makes the runtime an aggregation point for privilege.

The architecture developed in later chapters will give us better alternatives. For now, we need only preserve the distinction.

The runtime executes the actor. It is not automatically the actor.

The Agent

We can now describe the agent more precisely.

For the purposes of this architecture, an **agent** is a logical software actor assigned an objective and capable of selecting and initiating actions in pursuit of that objective within some authority context.

That definition is intentionally architectural rather than philosophical.

It does not require us to decide whether the agent genuinely understands its objective, whether its behavior constitutes reasoning in a human sense, or whether increasingly capable systems deserve some more expansive description of agency. None of those questions is necessary to design an enterprise authorization boundary.

The definition identifies the properties that matter to architecture.

The agent has an objective. It has some ability to select what to do. It can initiate actions.

And when those actions reach protected enterprise resources, it operates within an authority context.

Return to the procurement example.

The objective is to obtain one approved replacement laptop for Jesse Fox in Portland.

The agent can inspect relevant information and choose among possible next steps.

It can invoke capabilities that inspect inventory, retrieve purchasing standards, identify suppliers, prepare transactions, request approval, or submit an order.

Some of those capabilities merely return information. Others can change enterprise state.

The agent is the logical actor connecting those activities into pursuit of the assigned objective.

That makes it a useful unit of architecture.

An enterprise can assign an owner to it. It can classify the agent by risk. It can define which tools it may use. It can determine which kinds of tasks may be delegated to it. It can establish limits on its authority. It can evaluate the agent before granting additional autonomy. It can suspend the agent when necessary.

None of that requires pretending the agent is a person.

It requires recognizing it as an actor.

The Agent Instance

There is still one more distinction.

Suppose the procurement agent handles ten thousand requests during a month. Those requests may execute concurrently across many processes or containers. Instances may be created for individual tasks and destroyed afterward. Some tasks may run for seconds; others may pause while awaiting approval and resume later.

When a particular purchase order is questioned, knowing that the *Procurement Agent* created it may not be enough.

Which execution?

That is the role of the **agent instance**.

The logical agent identifies what kind of actor is operating. The instance identifies a particular execution of that actor where such distinction is required for attribution, isolation, credentialing, policy, or evidence.

The analogy to human identity is imperfect but useful.

Knowing that an employee is authorized to use an application is different from identifying a particular authenticated session through which an action occurred. The employee remains the same person across many sessions. The individual session provides additional context for a particular interaction.

Similarly, an agent may retain a stable logical identity while many instances execute tasks under that identity.

This distinction becomes especially important when authority is temporary.

An enterprise may decide that a particular agent instance should receive authority to perform one task for fifteen minutes. Another instance of the same logical agent should not automatically inherit that authority. If the first task completes, the enterprise may want the associated authority to disappear with it.

That design is difficult if the architecture recognizes only the long-lived agent.

It is nearly impossible if the architecture recognizes only the runtime.

The Transaction Is the Useful Unit of Analysis

The easiest way to expose those relationships is to stop looking at the agent as an isolated component and instead examine an entire agentic transaction.

Jesse Fox requests a replacement laptop.

That gives us a **principal.**

Some mechanism allows the procurement agent to perform appropriate work arising from that request.

That gives us **delegation.**

A particular logical actor and executing instance perform the work.

That gives us **agent identity**.

The work has an objective and boundaries.

That gives us the **task.**

The agent decides that it should purchase a particular device, but enterprise controls determine whether that proposed operation is permitted.

That gives us **policy.**

The agent reaches the procurement system through some exposed capability.

That gives us the **tool.**

The purchase request is submitted.

That gives us the **action.**

The enterprise has now committed funds and created an order.

That gives us the **outcome.**

Information connecting those events is preserved so the transaction can later be understood and reconstructed.

That gives us **evidence.**

We now have the complete architectural chain:

Principal → Delegation → Agent Identity → Task → Policy → Tool → Action → Outcome → Evidence

That is an important signal for architects, but it should not be overstated. As this book is being written, important parts of the agent identity and authorization landscape remain unsettled. NIST's NCCoE concept paper remains draft work. Industry protocols continue to evolve. Standards bodies are examining problems for which there is not yet one universally accepted architecture.

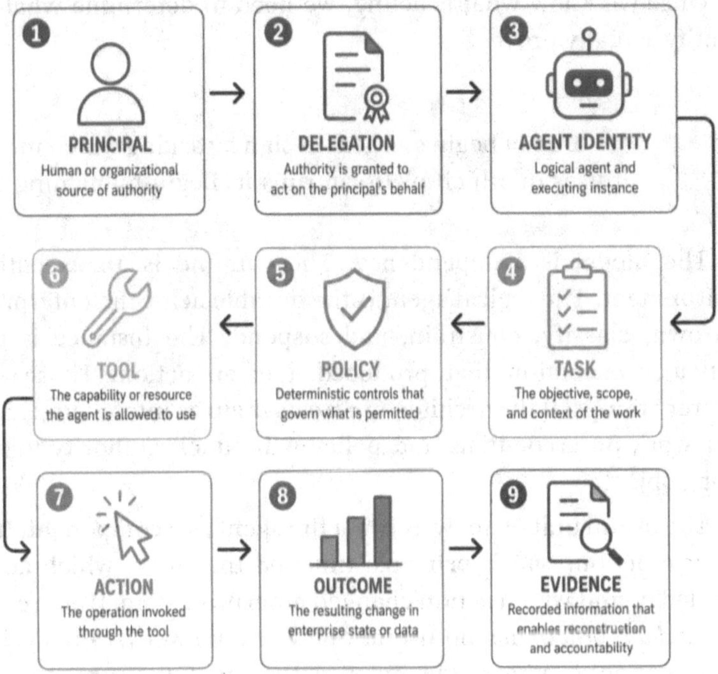

Agentic Transaction Anatomy
End-to-end flow of delegated action

The enterprise does not have the luxury of waiting for every question to be settled.

Agents are already being connected to enterprise systems.

The practical architectural response is to begin with distinctions we can defend.

The user is not the agent.

The model is not the agent.

The runtime is not the agent.

The logical agent is not necessarily the executing agent instance.

And none of those identities, by itself, tells us what the actor is authorized to do.

That last distinction takes us directly to the next problem.

Once we know what is acting, we need to determine what its identity actually proves.

Architect's Takeaway

Do not begin an agent design by asking which model it uses or which platform runs it. Begin by naming the actor.

The model is a dependency. The runtime is an execution environment. The logical agent is the durable actor the enterprise can own, classify, constrain, and suspend. The instance is the particular execution that produced a given action. Those are different things. If the architecture treats them as interchangeable, later work on credentials and policy will attach authority to the wrong object.

The useful unit of analysis is not the agent as a component. It is the transaction: which principal initiated the work, which actor pursued it, and which action changed enterprise state. *Who created the purchase order?* has no useful one-word answer. If the design review cannot separate the employee, the agent, the runtime, and the executing instance, it is not ready to discuss authentication.

The next question is different. Once the actor can be named, what evidence establishes that this software is that actor? That is Chapter 2.

Authority Check

- If this purchase order is questioned six months from now, which actor will the enterprise name?
- Is that name the model, the runtime, the logical agent, or a particular execution?

- Can two logical agents share a runtime without becoming the same actor?
- Can the same logical agent run ten thousand times without those executions becoming indistinguishable?
- Does the architecture treat a change of model as a change of actor?
- When the work pauses, migrates, or resumes, is the actor still the same thing the task was assigned to?

Decision Checklist

- The model is treated as a capability of the agent, not as the agent.
- The runtime is treated as the environment that executes the agent, not as the agent.
- The logical agent is the durable unit of ownership, classification, tooling, and suspension.
- An instance or execution identity exists where concurrent, asynchronous, or resumable work must be attributed.
- A consequential action can be described as a transaction connecting principal, actor, task, action, and outcome—not as "the AI did it."
- Identity questions that remain after the actor is named (binding, attestation, lifecycle, federation) are deferred to the identity chapter rather than smuggled into the definition of the actor.

- Can two logical agents share a runtime without becoming the same actor?
- Can the same logical agent run ten thousand times, with those executions becoming indistinguishable?
- Does the architecture treat a change of model as a change of actor?
- When the work pauses for hours or days and resumes, is the actor still the same thing that was delegated to?

- [illegible] the model is treated as a capability of the agent, not as [illegible].
- The runtime [illegible] that executes the agent, not is the agent.
- The logical agent is the durable unit of ownership, [illegible], and [illegible].
- [illegible] instance of execution [illegible] where concurrent, asynchronous, or resumable work must be distinguished.
- [illegible] can be described as a transaction [illegible] and as the AI [illegible].
- Identity [illegible] remain stable after the actor is [illegible]; changes are deferred to the [illegible] chapter rather than [illegible] into the [illegible] of the actor.

CHAPTER 2

Establishing Agent Identity

Chapter 1 identified the actor. That is not the same as proving the actor. A purchasing API that receives a request labeled Procurement Agent does not need another definition of agency. It needs evidence: that this is that agent, in an environment allowed to run it, as a particular execution of a particular task.

A logical agent exists long before any particular execution of it does.

It may be registered in an enterprise catalog, assigned an owner, associated with approved tools, deployed through several environments, and instantiated hundreds of times each day. Those executions may occur on different hosts, in different clusters, or even across different infrastructure providers. Some last seconds. Others wait for an event, resume after an approval, retry failed work, or survive the loss of the process that originally began the task.

That creates an identity problem ordinary application naming does not solve.

Suppose the enterprise has approved an agent called Procurement Agent. A request arrives at a purchasing API claiming to originate from it. The API needs evidence.

Is this actually Procurement Agent?

Is it executing in an environment in which Procurement Agent is allowed to run?

Is the calling process one of the workloads authorized to represent that agent?

If execution moved, resumed, or restarted, is this still part of the same task?

If several instances are operating simultaneously, which execution produced this request?

And if Procurement Agent and three other agents share the same orchestration platform, does the platform's identity establish the identity of any particular agent running inside it?

These are not authorization questions yet. The agent has not been granted authority. They are questions about a trustworthy relationship among the logical actor, the software executing it, and the particular execution that eventually produces an action.

That relationship is the subject of this chapter.

Start With What Workload Identity Already Knows

Enterprises do not need to invent machine identity because the software now contains an agent.

Modern workload-identity architectures already let software prove who it is from characteristics of its execution environment rather than from a password, API key, or other static secret embedded in configuration. The useful reference point is **SPIFFE**—the Secure Production Identity Framework for Everyone—and **SPIRE**, its common implementation. SPIFFE defines the model: an identity namespace, trust domains, workload attestation, short-lived credentials, and a Workload API so software can obtain and refresh that material without long-lived secrets in the application. A workload receives a SPIFFE ID and proves possession of it with a SPIFFE Verifiable Identity Document, or SVID. SPIRE attests the node and the process, compares what it sees with registered conditions, and issues the corresponding identity.

Neither was created for AI agents. That is why they are useful here. They solve a problem enterprises already understand: establishing the identity of software in dynamic infrastructure.

The architectural lesson matters more than the implementation.

Identity can be derived from trusted evidence about the execution environment rather than asserted by the software itself.

An agent should not be treated as Procurement Agent because its system prompt says so, because an orchestration file contains that name, or because the model introduces itself that way. Those are application-level assertions. They are not strong identity evidence.

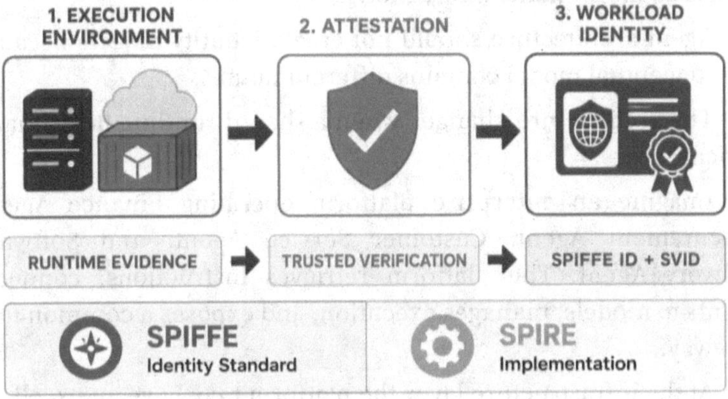

Workload Identity as the Foundation

The same applies to an agent identifier placed in an HTTP header by the caller. If any workload holding the platform credential can put any agent name in that header, the downstream resource has learned what the caller wants to be called. It has not established which agent is acting.

A trustworthy agent identity therefore needs a relationship to a trustworthy workload identity.

The challenge is deciding what that relationship should be.

Binding the Agent to the Workload

Consider a dedicated runtime first.

Procurement Agent runs as its own independently deployed service. It has its own deployment definition, runtime configuration, operational owner, and workload identity. No other logical agent executes in that workload.

In that architecture there is often little value in a second independently authenticated identity merely so that one credential says *Procurement Agent* while another says *the workload that runs Procurement Agent*. If the lifecycle and security boundaries are genuinely identical, the workload identity may represent the logical agent for many interactions.

Agent architecture should not create identity objects because the conceptual model contains different nouns.

The architecture changes when a shared runtime hosts many logical agents.

Imagine an enterprise platform operating Finance Agent, Procurement Agent, Customer Service Agent, and Software Delivery Agent. The platform retrieves instructions, connects agents to models, manages execution, and exposes a common tool gateway.

At the infrastructure layer the platform may have an excellent workload identity: attested hosts, short-lived credentials, strong service-to-service authentication, a controlled production environment.

None of that necessarily distinguishes Finance Agent from Software Delivery Agent.

If every downstream resource sees only `agent-platform-production`, the enterprise has authenticated the host and obscured the actor inside it.

The workload identity is a trustworthy base. Another binding is required if different logical agents sharing that workload must remain distinguishable.

There are several ways to accomplish this. The platform might obtain distinct credentials for the agents it is entitled to host. An identity service might issue an agent-specific assertion after validating the workload and the registration. A gateway might cryptographically bind platform identity, agent identifier, and execution context. The runtime might obtain different workload identities based on the agent configuration it has loaded.

A workload should be able to represent only those agent identities the enterprise has authorized it to host.

Without that binding, agent identity is self-asserted metadata. With it, the enterprise can say something stronger: this is an approved execution environment, and this environment has been permitted to instantiate this logical agent.

That still says nothing about what the agent may purchase, modify, disclose, or deploy. Those questions come later. But now we know what is running.

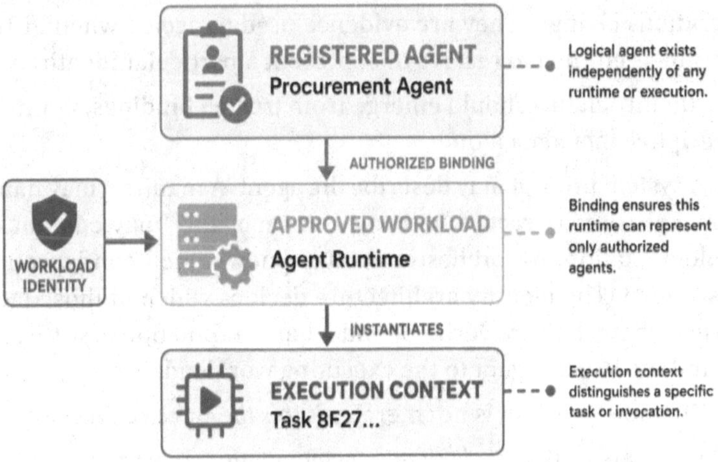

Binding Agent Identity to Runtime and Execution Context

Attestation Establishes the Base

Workload attestation is how the environment participates in the answer to *Who are you?* The process does not get to provide that answer alone.

SPIRE determines the identity of the caller by examining properties available from the execution environment and comparing them with registered conditions—operating-system characteristics, a Kubernetes service account or namespace, and similar facts.

The same principle applies one level up.

An agent platform may know that a configuration file says `procurement-agent`. Identity infrastructure should be interested in stronger facts. Which signed deployment is executing? Which namespace launched it? Which platform or cluster is hosting it? Which agent definitions is that runtime permitted to load? Is this production, test, or development? Has the registration been suspended?

Not every one of those facts belongs inside an identity credential. That would make the credential change whenever operations change. They are evidence used to decide whether the runtime is entitled to receive or represent a particular identity.

Identity claims should emerge from trusted bindings, not from descriptive metadata alone.

A system prompt may describe the agent. A manifest may name it. A registry may record its owner. A repository may contain its implementation. An orchestrator may know which configuration was loaded. The identity architecture decides which of those facts it trusts, how they are verified, and what combination is sufficient to bind the logical agent to the executing workload.

The new question is not merely: *Is this the expected process?*

It is also: *Which delegated actor is this process entitled to instantiate?*

Agent Identity Has a Lifecycle

Identities should follow a managed lifecycle rather than exist because someone once created a credential.

A logical agent begins with registration or provisioning. The enterprise establishes that it exists, identifies an owner, associates it with an intended environment, and records what is required to govern it. That source does not have to be a single enterprise-wide directory. Platform registries, configuration repositories, service catalogs, identity systems, and governance inventories can all serve. What matters is that the identity has an authoritative source.

Otherwise names proliferate faster than accountability. FinanceAssistant, finance-agent, fin-agent-prod, and AP-Helper-v2 may be one actor or four. A name embedded in source code is not an identity strategy.

The registry should establish a stable identity independent of incidental deployment details. Implementation, model, prompt, tools, and platform may change without creating a new logical actor.

Identity and version are different. If Procurement Agent moves from one model to another but retains the same purpose, ownership, governance, and security role, it is the same agent with a new version. If an experimental derivative has different ownership, access boundaries, or operational purpose, shared code is not a reason to share the identity.

Architects should be able to explain why two deployments represent the same logical agent or why they do not.

The lifecycle must also end. A retired agent should not obtain fresh identity material because an old deployment still exists. A suspended agent should stop receiving identities appropriate to production. A development copy should not become Procurement Agent in the production trust domain because someone duplicated its configuration.

Identity retirement is not housekeeping. It is part of the security boundary.

Identity Must Survive Runtime Discontinuity

Distributed systems have always had work that outlives the process that started it. Agents make this more visible because the unit of work can contain an evolving decision process rather than a predetermined sequence of calls.

A durable task may be persisted to a queue, workflow engine, state store, or agent platform. It may wait for a human response, schedule future work, delegate a subtask, or resume after infrastructure failure.

Execution identity and runtime identity therefore do not have identical lifetimes.

A robust design preserves enough information to establish that a resumed execution is legitimately continuing an existing task:

- the task still exists and is active;
- the logical agent is still permitted to execute it;
- the new runtime is permitted to host that agent;
- the resumed execution is associated with the correct persisted state;
- the prior execution has not already completed or been superseded.

Identity infrastructure does not have to enforce all of those conditions. Some belong to orchestration, policy, or task state. The architecture must preserve the identifiers and bindings those systems need.

Otherwise resume becomes recreation: a new process loads persisted state and announces that it is continuing an old task, with nothing trustworthy connecting the new execution to the old one.

That becomes dangerous once credentials and delegated authority are introduced. Chapter 3 takes up that problem. The identity lesson is enough here:

Durable work requires durable identity relationships even when the compute executing the work is ephemeral.

Containers are replaceable by design. The actor and its task may need to survive them.

Trust Domains Matter at the Boundaries

Inside one tightly managed platform, establishing agent identity can be straightforward. Enterprise architecture rarely stays that simple.

An agent may execute in one cloud and access resources in another. A corporate agent may invoke a partner service. An acquired company may keep its own identity infrastructure. A centralized platform may serve business units with distinct security boundaries.

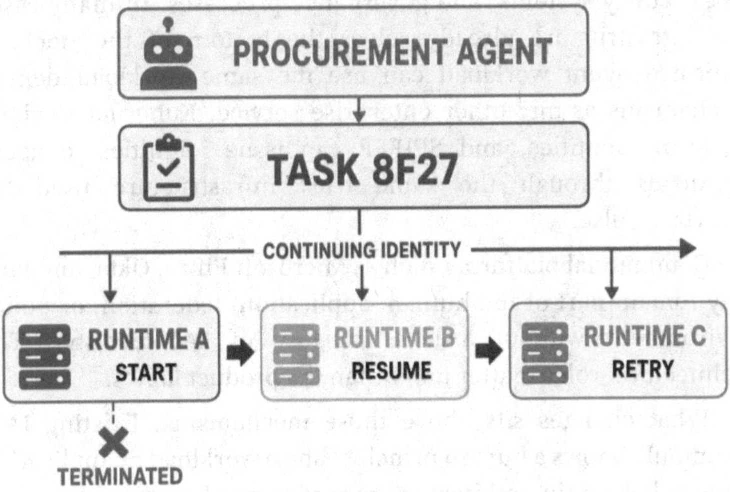

Identity Continuity Across Ephemeral Runtimes

SPIFFE already organizes identities into trust domains and supports federation through trust bundles. Federation does not make a foreign identity local. It establishes the cryptographic basis on which that identity can be recognized.

That is the right model for agents.

Crossing a boundary should not require the receiving system to pretend the agent belongs to its own identity domain. Recognition of a foreign identity should not imply permission to act.

Federation answers: *Do I trust the identity authority that says this is the actor?*

It does not answer: *What will I allow that actor to do here?*

Agent platforms often combine identity, orchestration, tools, and authorization into one developer experience. Architecture should pull those concerns back apart.

Fitting Agent Identity Onto Existing IAM

Most enterprises will not build a separate identity universe for agents, and they should not.

They already have identity providers, service principals, workload identities, certificates, secrets platforms, directories, access-policy systems, and governance processes. In many cases that infrastructure already solves the bottom of the stack. A dedicated agent workload can use the same workload-identity mechanisms as any other enterprise service. Kubernetes, cloud platform identities, and SPIFFE can issue identities to agent workloads through the same trust infrastructure used for everything else.

Commercial platforms such as Microsoft Entra, Okta, and Ping may remain part of the human, application, federation, or policy environment without a new category of "AI identity." The architectural roles matter more than the product labels.

What changes sits above those mechanisms. Existing IAM commonly knows a human principal and a workload or application principal. Agentic architecture may also need to preserve *which logical agent the workload is representing*, and, for consequential or long-running work, *which execution of that agent is responsible for the current task.*

Sometimes those distinctions can live inside existing identity infrastructure. Sometimes they should remain contextual claims bound to an authenticated workload. Sometimes the platform needs a registry or agent-identity service above conventional workload identity.

A logical agent does not automatically require a user account. An execution does not automatically require a service principal. A task does not automatically require an identity object. The enterprise needs trustworthy bindings, not an explosion of directory entries.

NIST's National Cybersecurity Center of Excellence is examining the same question in its February 2026 draft concept paper on software and AI-agent identity and authorization: how far existing workload and application identity approaches carry into agentic systems, and where additional treatment is required. As of this writing that paper is a draft, not a standard. The question is still the right one for architects.

Do not replace mature IAM because the workload contains an agent. Do not assume mature IAM has already modeled every distinction a delegated actor introduces.

When Identities Should Collapse—and When They Must Not

It is easy to overcorrect.

Once architects distinguish workload, agent, and execution identity, a design can acquire an elaborate model that contributes little control. The test is not whether the diagram contains separate boxes. The test is whether collapsing the identities destroys information needed for a security, operational, or accountability decision.

Collapse them when they share lifecycle, owner, trust boundary, policy treatment, and accountability. An internal summarization agent running as a dedicated service, with no consequential tools and ordinary request tracing, may need no second managed identity. A short synchronous execution may need only a trace identifier, not an independently authenticated instance.

Keep them apart when the opposite is true.

A shared runtime with one powerful platform identity is a warning when the agents inside it do different jobs. If a customer-service agent and a software-deployment agent are indistinguishable downstream, the platform identity is too coarse. The same problem appears when development and production can obtain the same identity; when a retired agent still receives credentials because the service principal was never disabled; when a long-running task resumes under a new process with no trustworthy link to the prior execution; and when a platform accepts an `agent_id` supplied by the caller without verifying that the authenticated workload may use it.

The problem is not a shortage of names. It is the absence of a trustworthy binding between the name and the thing executing.

That is the line to protect.

A Binding Model

The identity architecture can now be stated compactly.

A logical agent identity should be anchored in an authoritative registration and lifecycle.

An executing workload should establish its own identity through enterprise workload-identity mechanisms appropriate to the environment.

The enterprise should maintain a binding that establishes which workloads may instantiate or represent which logical agents.

Where execution-level distinction matters, the platform should create an execution identity or correlation context bound to the logical agent and task, able to survive replacement of transient compute.

When execution crosses trust domains, the receiving environment must establish which identity authority it trusts without confusing federation with permission to act.

Those relationships form a chain of identity evidence:

registered agent → approved workload → executing instance → continuing task

This is not the book's full Principal-to-Evidence chain in miniature. It answers a narrower question.

If a consequential request reaches an enterprise resource, can the enterprise establish that the software making the request is a legitimate execution of the logical agent it claims to represent?

If not, there is little value in debating how carefully the agent's permissions were designed.

The wrong actor may be exercising them.

Architect's Takeaway

Do not begin an agent-identity design by asking whether every agent needs a new service account. Chapter 1 named the actor. This chapter asks what evidence makes that name trustworthy.

Begin with the security boundaries existing workload identity does and does not distinguish. A dedicated agent workload may need no new identity mechanism at all. A shared runtime hosting materially different agents needs an explicit binding between the authenticated platform and the logical agent it is permitted to represent. Long-running or asynchronous work needs execution continuity that survives replacement of the underlying process. Cross-domain execution needs federation of identity without federation of authority.

The design-review question is therefore sharper than *How does the agent authenticate?*

Ask: *What evidence establishes that this runtime is entitled to represent this agent, and will that relationship remain trustworthy for the entire execution lifecycle?*

Identity answers *what is acting.* It does not answer *why it is allowed to.* That is Chapter 3.

Authority Check

- What is the authoritative source for the logical agent's identity and lifecycle?
- What evidence establishes that the current workload is permitted to represent that agent?
- Can a shared runtime claim the identity of an agent it was not intended to host?
- Can execution identity survive pause, retry, migration, and resume without being confused with transient process identity?
- Are production, development, retired, and suspended agents distinguishable at the identity boundary?

- When an agent crosses a trust domain, what establishes trust in its identity issuer?
- Can a retired or unauthorized runtime still obtain or present an active agent identity?

Decision Checklist

- Use existing enterprise workload identity as the foundation where it already solves the problem.
- Bind logical agent identity to workloads through verifiable platform or identity controls, not self-asserted names, prompts, or headers.
- Maintain an authoritative lifecycle for registration, activation, suspension, and retirement.
- Separate workload identity from agent identity when ownership, lifecycle, trust boundary, or accountability differ.
- Preserve execution or task identity when concurrent, asynchronous, resumable, or migrated work must remain distinguishable.
- Do not create separate security principals where a cheaper trustworthy binding already gives the required distinction.
- Treat cross-domain identity federation as recognition of an actor, not permission for that actor to act.
- Verify that retired, suspended, and nonproduction runtimes cannot obtain or represent active production agent identities.

CHAPTER

3

Delegating Authority

An identified agent is not yet an authorized agent.

The enterprise may know exactly which logical agent is executing, which workload hosts it, and which execution is responsible for the current task. The next question is different: *Why is this agent permitted to act?*

Authority has to come from somewhere. An employee may have authority to request a purchase. A manager may have authority to approve one. An organization may authorize certain classes of automated transactions. Whatever the source, an agent does not acquire legitimate authority merely because it has been deployed, registered, or authenticated.

Authority must reach it through delegation.

The architectural challenge is to move that authority without turning the agent into the principal and without handing the agent all of the principal's standing privileges.

The Principal Should Not Disappear

Return to thc procurement agent. Jesse Fox asks it to replace a failed laptop for the Portland office. The agent identifies an approved supplier, prepares the order, and eventually invokes a procurement API.

At the procurement system, two facts matter.

The first is that the procurement agent made the request.

The second is that it made the request under authority originating somewhere else.

If the procurement system records only the employee, the agent disappears from the transaction. The system appears to show that the employee directly submitted an API request that the employee may never have seen.

If it records only the agent, the source of authority disappears. The transaction may show that procurement-agent-17 created the order, but not why that agent was entitled to create it.

A delegated architecture needs both.

This is the essential difference between acting **as** a principal and acting **on behalf of** a principal.

Impersonation collapses the distinction. To the downstream system, the intermediary effectively becomes the subject whose authority it is exercising. That may be intentional in some enterprise designs, but it is a poor default for autonomous agents because it obscures precisely the relationship the architecture needs to preserve.

Delegation keeps the relationship visible: *This principal authorized this actor to exercise some authority on the principal's behalf.*

The actor remains the actor. The principal remains the principal.

Existing authorization standards already provide useful mechanisms for representing this distinction. OAuth 2.0 Token Exchange, standardized in RFC 8693, explicitly distinguishes impersonation from delegation. In its delegation model, a token can represent the subject while separately identifying the current actor through the act claim. The particular token format is less important than the architectural principle: *the identity exercising authority and the identity from which that authority originates should not have to collapse into one identity.*

For agents, that distinction is foundational.

Delegation Is an Authority Transfer, Not an Identity Transfer

Delegation does not mean giving the agent everything the principal can do.

Suppose Jesse Fox can create purchase requisitions, view departmental budgets, modify shipping addresses, approve certain expenses, access supplier information, and review previous transactions. Jesse asks the agent to order one approved laptop.

The task does not justify transferring that entire collection of privileges.

What needs to move is enough authority to perform the delegated task.

This gives us a useful relationship:

Principal authority ≠ delegated authority

The delegated authority derives its legitimacy from the principal without becoming identical to the principal's standing authority.

Human authorization is often durable. Employees accumulate roles, group memberships, application entitlements, and organizational permissions because they perform responsibilities over months or years. An agent execution may exist to accomplish one narrow task.

Its authority should reflect that difference.

A purchasing employee may need continuing access to procurement functions because purchasing is part of the employee's job. An agent created to fulfill one request does not necessarily need continuing purchasing authority after that request is complete.

The agent's authority should follow the task.

The Task Is Part of the Delegation

An authenticated procurement agent may be permitted to call the procurement API. That still does not tell us whether the current invocation belongs to an authorized task.

The task supplies that context.

An employee might initiate:

Purchase one approved laptop for the Portland office, billed to CC-4410, not to exceed $4,000.

Architecturally, the resulting delegation might establish:

- the originating principal;
- the delegated actor;
- the task for which authority is being granted;
- the protected resource or audience;
- the permitted kind of action; and
- the period during which the authority is valid.

This does not mean stuffing a natural-language prompt into an access token. The task may be represented elsewhere and correlated through identifiers or authorization context. What matters is that the authority exercised by the agent can be connected to the work that justified its issuance.

That becomes particularly important when the same logical agent performs different work for different principals. A procurement agent may execute hundreds of tasks during a day. One may involve purchasing office equipment. Another may involve researching an existing order. A third may involve preparing a transaction that someone else must approve.

The logical agent is the same.

Its delegated authority need not be.

The enterprise therefore needs a relationship that workload identity alone cannot establish:

Principal → Task → Delegated Actor

The credential used during execution should represent or derive from that relationship rather than merely proving that the agent exists.

Two Convenient Failure Modes

The simplest approach is often to give the agent a powerful service credential.

The procurement agent receives a service principal, service account, workload identity, API key, or equivalent credential with permission to create purchase orders. Whenever the agent needs to create one, it uses that credential.

Operationally, this is convenient.

Architecturally, it leaves an important question unanswered: *Why is the agent creating this purchase order?*

The credential proves that the agent is generally permitted to create orders. It does not necessarily establish which principal initiated the work, whether that principal was entitled to initiate this transaction, or whether the credential is being exercised within the task that justified the action.

The agent has standing authority independent of the immediate delegation.

For conventional deterministic services, broad service authority has long been an accepted design tradeoff. Application logic determines which operations are appropriate, and the enterprise trusts that implementation to use its privilege correctly.

That assumption deserves greater scrutiny when the actor selecting among actions is probabilistic. The credential may authorize a larger action space than the current task requires while the component selecting within that action space is no longer completely predetermined by application code.

The opposite shortcut is user-token passthrough.

Instead of giving the agent broad standing authority, the application passes the user's credential downstream. The agent can now exercise whatever authority that credential conveys.

This creates a different problem. The agent may disappear as a distinct delegated actor, and the user's existing authorization envelope may become the agent's authorization envelope even though the user accumulated those privileges for many unrelated responsibilities.

The two shortcuts fail differently.

With a broad service credential, the agent possesses authority because of what the agent is generally allowed to do.

With user-token passthrough, the agent possesses authority because of what the user is generally allowed to do.

Neither necessarily expresses what this agent is allowed to do for this principal for this task.

That is the delegation problem.

Task-Bound Authority

A stronger architecture creates authority in response to the task rather than treating authority as a permanent property of the agent.

The mechanism will vary. OAuth-based environments may issue or exchange access tokens. Other environments may use temporary credentials, capabilities, signed authorization artifacts, or platform-specific security contexts.

The mechanism is secondary to the properties of the resulting authority.

The delegated authority should remain attributable to the originating principal. It should identify the actor exercising it where that distinction is required. It should be directed toward the intended resource or audience. It should permit the appropriate kind of action. It should expire on a timescale appropriate to the task.

This is task-bound authority.

OAuth provides useful building blocks for parts of this architecture. RFC 8693 defines token exchange and provides semantics for delegation and impersonation. RFC 9396 defines Rich Authorization Requests, allowing structured authorization requirements to be expressed through authorization_details rather than relying exclusively on broad scope strings.

Neither specification defines an authorization architecture for autonomous AI agents. They are mechanisms that an agent architecture can use.

Current standards work is still developing. NIST's software and AI-agent identity and authorization work is examining how existing and emerging identity technologies apply to agents and where additional mechanisms may be needed. Emerging IETF work is exploring related questions. These efforts are important research inputs, but proposals and concept work should not be mistaken for settled standards.

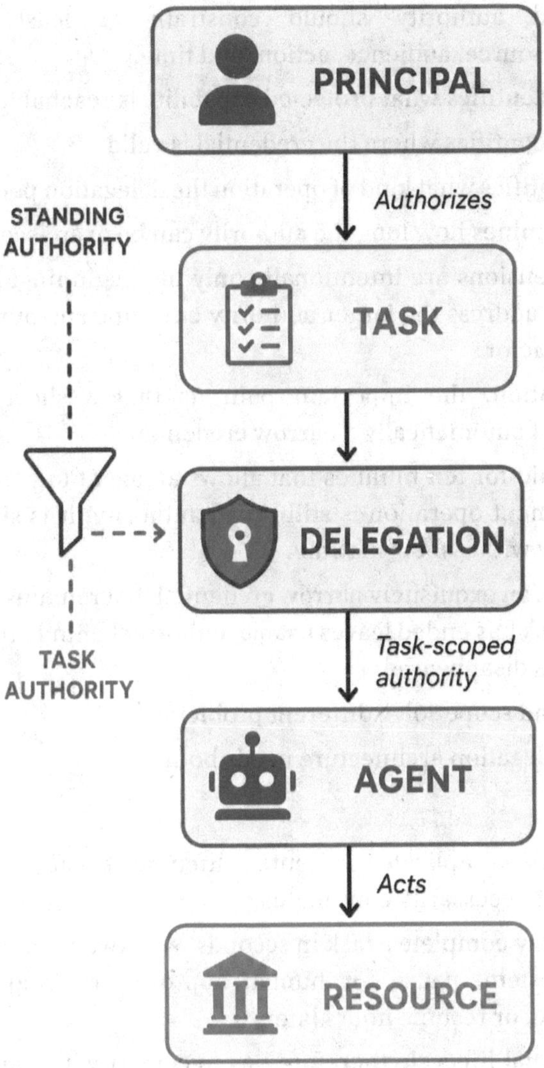

Task-Bound Delegation

The durable architectural principle does not depend on how that work develops. Authority granted to an agent should correspond to the delegated task, not merely to the standing privileges of either the principal or the agent.

Scope and Lifetime Solve Different Problems

Task-bound authority should constrain at least four dimensions: resource, audience, action, and time.

Resource identifies what protected capability is reachable.

Audience identifies where the credential is valid.

Action identifies what kind of operation the delegation permits.

Time determines how long the authority can be exercised.

These dimensions are intentionally only the beginning. Later discussion will address the larger authority envelope surrounding a probabilistic actor.

For delegation, the important point is that a short-lived credential is not automatically a narrow credential.

A token valid for ten minutes that allows an agent to perform every procurement operation is still broad authority. It is simply broad authority with a short lifetime.

Conversely, an exquisitely narrow credential that remains valid long after its task has ended leaves usable authority behind after its justification has disappeared.

Duration and scope solve different problems.

A sound delegation architecture needs both.

Credential Lifetime Should Follow Task Lifetime

Agent tasks complicate credential lifetime because their execution is not necessarily continuous.

An agent may complete a task in seconds, wait twenty minutes for another system, pause for human approval, retry after a transient failure, or resume hours later.

The credential lifecycle therefore has to coexist with the task lifecycle.

If the task finishes, authority associated only with that task should cease to be useful.

If a credential expires while legitimate work remains, the architecture needs a controlled way to reacquire appropriate authority rather than silently falling back to a broader credential.

If execution pauses and resumes, the system should determine whether the original delegation remains valid under current conditions.

If work moves to another runtime after a failure, a transient credential should not simply become a portable secret that any replacement workload can replay.

These are not token-management trivia. They determine whether task-bound authority remains task-bound once the system encounters normal distributed-system behavior.

Delegation Has a Control-Plane Cost

There is a reason enterprises gravitate toward long-lived service credentials: they are operationally simple.

Task-bound authority is not.

A large agent fleet may create enormous numbers of short-duration executions. If tasks require credential issuance, exchange, renewal, or reauthorization, identity and authorization infrastructure becomes part of the execution control plane.

Issuance volume matters. Latency matters. Authorization-service availability matters. Clock skew matters. Credential renewal matters. Audit volume matters.

Long-running tasks create expiration and reauthorization problems. Retries create questions about reuse versus reissuance. Asynchronous workflows may need to retain enough delegation context to obtain fresh credentials without persisting powerful bearer credentials.

Caching requires particular care. A credential issued for one task but cached at the logical-agent level and reused across subsequent tasks has quietly stopped being task-bound. An optimization has recreated standing authority.

The architecture should therefore distinguish caching delegation context from caching usable authority.

A system may preserve enough information to establish that a task was legitimately delegated and obtain a fresh credential when necessary. That is materially different from retaining a live credential that can continue exercising authority across tasks.

The goal is not maximum credential churn. It is alignment between the lifetime of authority and the lifetime of the work that authority exists to support.

Token Exchange as an Architectural Pattern

The identity infrastructure established earlier gives the enterprise trusted principals and actors. Delegation now requires a way to derive authority appropriate to a particular task rather than simply reusing either party's standing credentials.

Token exchange demonstrates one useful pattern.

Suppose authority associated with an initiating principal reaches an agent that needs access to the procurement API.

Instead of forwarding the original credential unchanged, an authorization service can evaluate the delegation and issue a different credential appropriate to the downstream resource.

Conceptually:

Upstream authority → Delegation evaluation → Task-appropriate credential

The resulting credential can differ in audience, permitted action, lifetime, and other authorization characteristics. Where appropriate, it can preserve the distinction between the subject of the authority and the actor exercising it.

The important architectural event is not the token exchange itself.

It is the transformation of authority.

An authorization server that exchanges a broad user token for an equally broad agent token has added protocol machinery without materially improving the authority model.

A useful exchange determines what upstream authority should survive, what should be removed, and what additional constraints should apply to the downstream actor.

The same test applies to non-OAuth mechanisms.

Whenever upstream authority becomes downstream authority, architects should ask: *What was preserved, what was removed, and what was constrained?*

Delegation That Continues Downstream

An agent authorized to perform a task does not automatically possess authority to delegate that task. Acting and delegating are separate privileges. Without that distinction, an agent can convert narrow execution authority into a broad chain of actors that the principal never selected and the enterprise cannot reliably control.

When Agent A assigns part of its work to Agent B, the downstream grant should identify the originating principal, the parent task, the specific subtask, the receiving agent, and the authority conveyed. Agent B should receive only what its subtask requires. It should not inherit Agent A's complete credential, tool set, data access, or remaining transaction ceiling.

Authority should attenuate across the chain. A downstream grant may be narrower in action, resource, amount, duration, data scope, or ability to delegate again, but it should not become broader than its parent. If Agent B requires additional authority, that increase must come from a principal or policy decision empowered to grant it—not from the fact that another agent requested help.

The task relationship must also survive independently of credentials. Short-lived tokens may be exchanged or renewed as execution proceeds, but the delegation record should preserve why Agent B acted, which authority supported the action, and how the result returned to Agent A's task. This produces a delegation chain rather than a collection of unrelated workload calls.

Each hop is therefore an independently authorized transition. The enforcement system should evaluate whether the sending agent may delegate, whether the receiving agent may accept the task, whether the proposed authority is properly attenuated, and whether organizational or trust-domain boundaries require additional approval. Emerging agent-specific protocols may standardize portions of this exchange, but the durable requirement is preservation of principal, task, attenuation, and evidence across every hop.

The Delegation Record Must Outlive the Credential

Credentials are transient. Accountability is not.

A task may require several credentials for several systems. Individual credentials may expire long before anyone investigates the resulting transaction.

The credential therefore cannot be the enterprise's only record that delegation occurred.

The later evidence architecture will connect principal, delegation, task, actor, policy decisions, actions, and outcomes. For now, the requirement is narrower:

Credential issuance should be attributable to a delegation event that survives the credential's operational lifetime.

The enterprise should be able to establish why authority was issued, to which actor, for which task, under which relevant constraints, and through which authorization decision.

The credential exercises the authority.

The delegation record explains where the authority came from.

Task-Bound Delegation

The resulting architecture does not depend on a particular identity product or token format.

A principal initiates or authorizes a task. The enterprise determines that the principal may delegate the relevant work. An identified agent is selected to perform it. A delegation service or equivalent authorization function derives appropriate temporary authority from legitimate upstream authority. The agent execution receives a credential directed toward the intended protected resource. The resource receives sufficient identity and delegation context to distinguish the relevant principal from the actor where necessary. The credential expires or otherwise ceases to be useful as its task context ends. The delegation event remains available for later reconstruction.

Authority originates with the principal but is transformed into temporary, task-appropriate authority for the identified agent. The agent neither becomes the principal nor inherits the principal's complete standing privileges.

Earlier discussion established which actor is executing. Delegation establishes where its authority came from and why it possesses that authority for this task.

That still leaves a consequential question unanswered.

A delegation can be completely legitimate and still grant too much authority.

The next problem is determining how much authority a probabilistic actor should receive.

Architect's Takeaway

In design review, do not accept "the agent uses the user's token" or "the agent has a service account" as a complete authority architecture. Both describe how software obtains usable credentials. Neither necessarily establishes the delegation under which a particular consequential action occurs.

Require the architecture to preserve the distinction among principal, delegated actor, and task. Authority should be derived for the work being performed rather than inherited wholesale from either the principal's standing privileges or the agent's application identity. Short-lived credentials help, but lifetime alone is not enough; the authority must also be appropriately directed and constrained.

Task-bound delegation adds control-plane complexity. That complexity should be acknowledged rather than designed away through standing privilege. The alternative may be operationally simpler while making it substantially harder to answer why an agent possessed the authority it exercised.

Authority Check

- Can the system identify both the originating principal and the agent exercising delegated authority?
- Does the agent act on behalf of the principal, or does the implementation effectively impersonate the principal?
- Is authority issued because of a particular task, or does the agent rely primarily on standing privileges?
- Can a credential issued for one task be reused for another?
- What happens when a legitimate task outlives its credential?
- Can the enterprise later establish why the credential was issued?

Decision Checklist

- Delegated actions remain attributable to both principal and acting agent where required.
- Credential caching does not convert task-bound authority into standing authority.
- Agent credentials do not automatically inherit the principal's complete standing authority.
- Credentials are associated with an authorized task or equivalent delegation context.
- Resource and audience restrictions prevent unintended credential reuse.
- Credential lifetime is aligned with task lifetime, including controlled renewal where necessary.
- Delegation and credential-issuance records survive long enough to support reconstruction.

CHAPTER 4

Bounding Agent Authority

Traditional least privilege begins with a reasonably stable assumption: we know what the software is going to do.

A payroll service reads payroll data and submits payments. A reporting service queries a defined set of tables. A deployment pipeline reads artifacts, modifies infrastructure, and invokes deployment operations according to a predetermined workflow. Even when the software is complicated, its possible behaviors are largely products of code paths that architects and engineers can inspect before execution.

An agent changes that assumption.

The enterprise may know the agent's objective without knowing in advance which sequence of actions it will select to accomplish it. The agent may choose among tools, retrieve different information depending on what it discovers, revise its plan, repeat operations, abandon one approach for another, or decide that an intermediate action is necessary even though nobody explicitly enumerated that action when the task began.

That flexibility is much of the reason to use an agent. It is also why ordinary formulations of least privilege become harder to apply.

NIST has identified precisely this unresolved question in its 2026 concept work on software and AI-agent identity and authorization: how should least privilege be established when an agent's required actions may not be fully predictable at deployment time? The question remains part of active standards and architecture work rather than a solved protocol problem.

The answer cannot be to predict every decision the agent will make. If that were possible, much of the reason for giving the agent discretion would disappear.

The architecture instead has to bound the space within which those decisions may have consequences.

That space is the agent authority envelope.

Least Privilege For Nondeterministic Flows

Least privilege is often implemented by enumerating permissions: this identity may read these records, invoke these operations, write to these locations, and nothing else.

That remains necessary for agents, but it is no longer sufficient as the complete design model.

Suppose the procurement agent introduced earlier has been delegated Jesse Fox's request: one approved replacement laptop for the Portland office. We have already established how authority can reach that agent without simply handing it the principal's standing credentials. But even properly delegated authority can still be much too broad.

The agent might be able to search the approved catalog, read inventory, request quotations, create purchase orders, modify existing orders, select shipping destinations, add vendors, change payment information, cancel purchases, or approve invoices. Perhaps the principal who initiated the task legitimately possesses several of those privileges.

The agent does not necessarily need them all.

More importantly, "may create purchase orders" is itself too coarse a description of appropriate authority.

For this task, the meaningful boundary might instead be:

The agent may purchase one approved laptop from the approved vendor, for the Portland office, against cost center CC-4410, at $3,850, not to exceed $4,000, during the next four hours. It may create that order. It may not modify vendor banking information, create a new vendor, alter payment terms, purchase a second machine, or exceed the task ceiling.

That is considerably closer to the authority the task actually requires. An agent can make an entirely legitimate authentication request, possess legitimately delegated credentials, invoke an approved enterprise API, and still create an unacceptable consequence if the authority represented by those credentials is broader than the task.

OWASP describes excessive agency in closely related terms: excessive functionality, excessive permissions, and excessive autonomy increase the range of damaging actions an LLM-based system can perform when its output is mistaken, ambiguous, or manipulated. Its guidance consequently recommends minimizing available functions and downstream permissions rather than depending on the model to refrain from using authority it does not need.

The architectural objective is therefore not merely to reduce what an agent *usually does*. It is to reduce what the enterprise will permit the agent to cause.

That is a different problem.

The Authority Envelope

An authority envelope defines the outer limits of legitimate consequence for a particular agent operating under a particular task or class of tasks.

It should be thought of as multidimensional rather than as a role or flat permission list.

An envelope can constrain at least six dimensions:

Actions. What kinds of operations may the agent perform?

Resources. Against which systems, records, accounts, repositories, or other objects may those actions be performed?

Data. Which information may the agent read, combine, retain, disclose, or use as input to subsequent actions?

Context and time. Under which temporal, environmental, task, tenant, user, location, workflow, or other contextual conditions does the authority remain valid?

Consequence. What financial, operational, administrative, or externally visible effect may the agent create?

Consumption. How much compute, model usage, paid API capacity, tool usage, storage, or other metered resource may the agent consume?

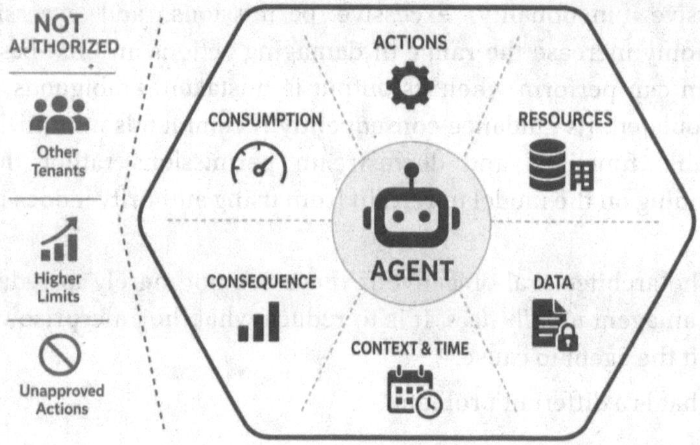

Agent Authority Envelope

Constrains actions, resources, data, context, consequence and consumption without attempting to predetermine every decision the agent will make.

These dimensions interact.

An agent authorized to issue refunds may be limited to customers assigned to a particular business unit. It may issue refunds below $250 automatically but only prepare larger refunds for approval. It may process no more than twenty refunds in an hour. Its authority may expire when the customer-service case closes. It may be prohibited from issuing a refund to a payment destination other than the one used for the original transaction.

None of those restrictions requires predicting which customer the agent will encounter or exactly how it will reason about the case.

They bound the consequences of that reasoning.

This is the essential adaptation of least privilege for probabilistic actors: *Do not attempt to make the permitted reasoning space deterministic. Make the consequential action space bounded.*

The agent may retain substantial freedom inside the envelope. Outside it, the enterprise should not depend upon the agent deciding to stop itself.

How that boundary is technically enforced is the subject of later chapters. At this point the important design decision is defining the envelope itself.

Permission Is Not Binary

One reason agent authority becomes overbroad is that enterprise permissions are frequently modeled around verbs such as read and write.

Agentic systems need a more consequential vocabulary.

Consider a software-development agent. "Write repository" might technically permit the agent to modify a source file, rewrite a build pipeline, change an authorization policy, alter deployment configuration, remove tests, modify ownership metadata, or commit directly to a protected branch.

Those operations are all writes. Architecturally, they are not equivalent.

A more useful progression is:

Read – observe protected state without modifying it.

Recommend – propose what should be changed.

Prepare – construct the intended change or transaction without causing the final consequence.

Approve – exercise authority over another proposed action.

Execute – create the consequence in the target system.

This progression should not be mistaken for a universal five-level maturity model. Different domains will require different verbs. The important point is that permissions should reflect consequential distinctions that matter to the business.

A procurement agent that may prepare a purchase order but not submit it possesses meaningfully less authority than one that can execute the purchase.

A coding agent that may create a pull request possesses less consequential authority than one that may merge to the protected branch.

An infrastructure agent that may generate a Terraform plan possesses less authority than one that may apply it.

A customer-service agent that may calculate a refund possesses less authority than one that may initiate payment.

These distinctions allow enterprises to preserve much of the value of agent reasoning without granting equivalent freedom over consequence.

They also provide natural points for progressive authority.

Progressive Authority

There is no architectural requirement that an agent receive its maximum possible authority when execution begins.

Authority can expand as the task develops.

A procurement agent might initially receive only catalog and inventory access. Once it identifies an appropriate product, it may receive authority to obtain a quotation. Once price and vendor conditions satisfy policy, it may receive authority to prepare an order. Only after additional conditions are satisfied might the architecture permit execution.

This is not simply human approval inserted between steps. Human approval is one possible mechanism and receives separate treatment later in this book. The broader principle is that authority can be staged.

An agent should receive consequential authority when that authority becomes necessary, not merely because the task might eventually require it.

This is particularly valuable when agents perform exploratory work. Research, planning, retrieval, comparison, and preparation may occupy most of an agent's execution while requiring little or no authority to modify enterprise state. Granting write or transactional authority throughout that entire period creates exposure without corresponding benefit.

Progressive authority reduces the duration as well as the scope of that exposure.

It also changes the architecture review question. Instead of asking, "What permissions does this agent need?" architects should ask: *At what point in the task does the agent need each kind of authority,*

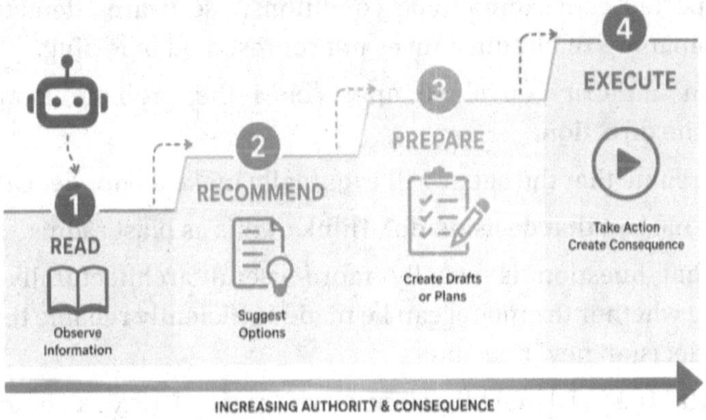

Progressive Authority

Authority is only granted when needed

and what conditions must be true before that authority becomes available?

That question often produces a substantially narrower design.

Reduce Consequence, Not Merely Error

Agent safety discussions frequently concentrate on making the agent less likely to make a mistake.

Better models may improve reasoning. Better instructions may reduce ambiguity. Better retrieval may improve factual grounding. Evaluation may expose problematic behavior. Additional context may help the agent choose correctly.

All of those measures can matter.

None substitutes for limiting authority.

There is an important difference between reducing the *probability* of an incorrect decision and reducing the *consequence* of one.

If an agent has unrestricted authority to delete production resources, reducing its erroneous deletion rate from one hypothetical value to a much smaller one does not change the maximum consequence when the error occurs. Nor does it adequately address malicious inputs, compromised tools, unexpected environmental conditions, software defects, or combinations of circumstances not represented in testing.

An authority envelope approaches the problem from the opposite direction.

Assume that the agent will eventually make a poor decision.

What can that decision do? Think of this as blast radius.

That question is usually more useful architecturally than asking whether the model can be made sufficiently reliable that the poor decision never occurs.

NIST's AI risk-management work treats AI risk as something that must be governed across the system lifecycle rather than eliminated solely through model performance, while OWASP's guidance on excessive agency explicitly recommends restricting functionality and permissions and using rate limits to constrain damage.

For delegated agency, this leads to a durable rule:

Model quality affects the probability of failure. Authority design affects the possible consequence of failure.

An enterprise needs both.

Only one of them defines the blast radius.

Transaction Ceilings Are Authorization

Financial systems make this especially obvious.

Imagine an accounts-payable agent with authority to create payments. A traditional permission model might ask whether the agent is permitted to invoke CreatePayment.

An authority-envelope design asks substantially more.

- For which legal entity?
- Against which accounts?
- To which classes of payees?
- For which invoice types?
- Up to what amount per transaction?
- Up to what cumulative amount per task or day?
- May the destination account differ from the vendor master record?
- May the agent create or modify the vendor whose invoice it is paying?
- May it split a transaction into multiple smaller payments?
- Does a higher amount require a different authority transition?

The same principle applies outside finance.

A cloud-operations agent may create infrastructure, but perhaps no more than a defined number of instances within a particular subscription and resource group.

A communications agent may send outbound messages, but perhaps no more than 500 recipients for the assigned campaign.

A SaaS administrative agent may modify tenant configuration, but only for the tenant associated with the task.

A repository agent may modify source files but not branch protections, secret stores, identity configuration, or deployment policies.

Transaction limits are therefore not merely operational safeguards added after authorization. They are part of the definition of what has been authorized.

A request for $500 and a request for $500,000 may invoke the same API operation. They do not represent the same authority.

Consumption Is Also Authority

The same reasoning applies to resources that do not look like traditional protected business transactions.

Agents consume things.

They invoke models. They call paid APIs. They provision compute. They query data platforms. They allocate storage. They execute searches. They retry operations. They may recursively invoke tools or other capabilities. In sufficiently autonomous systems, they may continue consuming resources while pursuing a goal long after a conventional application would have reached a fixed execution boundary.

That consumption is frequently treated as a FinOps concern.

For an agent, it is also an authority concern.

If the enterprise delegates authority to use a paid model, allocate cloud infrastructure, invoke a commercial data service, or execute a metered tool, it has delegated authority to consume organizational resources.

The authority envelope should therefore be capable of expressing consumption boundaries such as:

- maximum model expenditure per task;
- maximum number of model or tool calls;
- maximum compute allocation;
- maximum retries;
- maximum execution duration;
- maximum paid API consumption;
- maximum downstream work initiated.

OWASP's current AI Agent Security guidance explicitly recommends token, cost, retry, and tool-chain limits and identifies unbounded API or compute consumption as a denial-of-wallet risk.

The architectural significance extends beyond cost control.

Consumption boundaries also limit runaway behavior.

An agent caught in a planning loop is less consequential if it can make twenty tool calls than if it can make twenty thousand. An infrastructure agent that may provision four instances presents a different operational risk from one that may provision four thousand. A research agent authorized to purchase $50 of external data access has a different authority envelope from one backed by an unrestricted organizational account.

Resource consumption should therefore be included when architects ask how much authority an agent possesses.

The bill is sometimes the evidence that authority was exercised.

Data Scope Can Expand Consequence

Authority over data is similarly more complex than whether an agent has read permission.

Agents aggregate.

An individual record may be appropriately accessible. So may a second record. The combination of information across sources, users, tenants, time periods, or sensitivity categories may create a dataset with consequences greater than those of any individual retrieval.

NIST's 2026 agent identity and authorization concept paper explicitly raises this issue, asking how authorization should respond when an agent gains access to new tools and resources and aggregates information whose combined sensitivity may differ from that of its individual sources.

That means the authority envelope may need to constrain not merely resources but relationships among resources.

A customer-service agent might legitimately read the customer record associated with the active case. That does not imply authority to search every customer record for comparison.

A multitenant administrative agent may possess the technical capability to query a shared data store. Its authority envelope still belongs to one tenant.

A financial-analysis agent may access several datasets individually but lack authority to export their joined result.

Soon, we will examine data, context, tenant, retrieval, and execution boundaries in detail. The point here is narrower: the data an agent can accumulate determines part of what the agent can subsequently do.

Data access therefore contributes to the authority envelope even before the agent changes enterprise state.

Memory Can Become Standing Authority

Persistent agent memory introduces a less obvious version of the same problem.

Memory is often discussed as though it were simply additional context: information retained so that the agent can work more effectively in a later interaction.

Some memory is exactly that.

But persistent state can also carry authority-relevant information across task boundaries.

Suppose an agent remembers that a manager approved a particular vendor exception last month. Does that approval apply to a new purchase?

Suppose it retains a previously resolved account identifier, a deployment target, an approval state, a task commitment, a customer relationship, or an instruction granting unusual discretion. When the original task ends, what causes that information to stop influencing future action?

The risk is not that memory literally becomes an access token. The architectural problem is subtler.

Persistent memory can allow the *effect* of prior authority to survive after the conditions that justified it have disappeared.

An agent may reason from remembered approval as though approval remains current. It may retain a commitment created under one principal and apply it under another. It may carry sensitive context into a later task. It may treat a previously permitted exception as a standing rule.

OWASP's current agent-security guidance identifies memory poisoning and cross-session memory exposure as agent-specific risks and recommends isolation and validation of retained memory.

From an authority perspective, the governing principle is:

Task authority should expire even when task context does not.

Persistent memory therefore needs classification.

Some retained state is informational. Some is sensitive. Some is authority-relevant. Some should expire with the task. Some may legitimately persist but must be revalidated before it can influence consequential action.

An architecture that carefully expires credentials while allowing remembered approvals, permissions, commitments, or exceptional instructions to persist indefinitely has solved only the mechanical half of authority expiration.

Context Narrows Authority

The same action can be permissible in one context and impermissible in another.

An agent may be authorized to modify a customer account while handling an authenticated support case, but not later simply because it remembers the account.

It may be permitted to deploy during an approved change window but not after that window closes.

It may be permitted to purchase replacement equipment for a specific cost center but not transfer that purchasing authority to another department.

It may be permitted to access data while operating inside one tenant context but not after switching tenants.

This makes context a first-class part of the authority envelope.

Traditional authorization systems already provide many of the underlying concepts. Zero-trust architecture, for example, rejects implicit trust based on network location and emphasizes resource-oriented access decisions based on authenticated subjects and applicable conditions. NIST SP 800-207 was finalized in August 2020 and provides a mature foundation for that resource-centric approach.

Agents add a further requirement: relevant context may change during execution because the agent itself changes what it is doing.

A long-running agent may move from research to preparation to execution. It may discover a new resource. It may change tools. It may encounter data of greater sensitivity. It may reach a financial threshold. It may enter a portion of the task that carries substantially greater consequence.

The authority envelope should not be assumed to remain appropriate merely because the task has not technically ended.

Authority can contract, expire, or require transition as context changes.

Authority Envelopes Should Be Task-Shaped

Enterprise access control traditionally tends to accumulate around durable identities and roles.

Employees join groups. Applications receive service identities. Roles accumulate permissions. Exceptions become permanent. Over time, the resulting access represents what an identity might reasonably need across many possible activities.

Agent authority should push the architecture in the opposite direction.

The useful question is not: *What may this agent generally do?*

It is: *What may this agent do for this task, under these conditions, with these limits?*

That does not mean every permission must be constructed from scratch for every invocation. Enterprises will need reusable authority profiles, policy templates, capability sets, risk classes, and other abstractions.

But those abstractions should resolve into an authority envelope appropriate to the current work.

The distinction is important because a logical agent may perform tasks with very different risk.

A software-engineering agent reading a repository to explain a defect requires one authority envelope.

The same agent preparing a code change requires another.

Merging that change requires another.

Deploying it to production requires another still.

The logical identity did not change. The task did. Therefore the appropriate authority changed.

The Envelope Is a Maximum, Not a Goal

An authority envelope defines what an agent is permitted to do, not what it is expected to do.

That distinction is easy to overlook.

If the envelope allows purchases up to $4,000, the agent should not interpret $4,000 as a target. Jesse's approved item is $3,850. That is the order. The remaining $150 is not a budget to spend. If it permits 100 API calls, the agent need not make 100. If it permits access to five repositories, the task may require only one.

The envelope describes maximum legitimate authority. Reasoning still determines what action appears appropriate within it.

Later chapters will establish deterministic enforcement around those proposed actions. For now, the crucial architectural separation is between *decision freedom* and *consequence freedom*.

An enterprise can permit substantial decision freedom while tightly limiting consequence freedom.

That is the key to useful bounded agency.

The goal is not to make an agent behave like a deterministic workflow. It is to let the agent exploit its ability to reason, plan, adapt, and choose while preventing that flexibility from becoming unrestricted organizational authority.

Architect's Takeaway

Least privilege for agents cannot be reduced to assigning a smaller role to an AI service account. The architecture must describe the maximum consequential space within which a probabilistic actor may operate: which actions, against which resources and data, under which conditions, for how long, with what transactional and consumption limits.

The design-review question should therefore change from *"Does the agent have the permissions it needs?"* to *"What is the maximum consequence this authority permits if the agent makes the worst plausible decision available inside it?"* That question exposes overbroad tools, standing permissions, excessive transaction limits, persistent state, unrestricted consumption, and other authority that ordinary permission reviews can miss.

Authority Check

- Does the agent's authority reflect its current task, or the accumulated privileges of its principal, runtime, or logical agent?
- Which actions can the agent execute directly, and which can it only read, recommend, or prepare?
- What is the maximum financial, operational, administrative, data, or external consequence the authority envelope permits?
- Which contextual changes should narrow, expire, or transition the agent's authority?
- Can persistent memory carry approvals, commitments, permissions, or other authority-relevant state beyond the task that justified them?

- What organizational resources can the agent consume, and are those consumption rights explicitly bounded?

Decision Checklist

- The maximum consequence of an erroneous but technically authorized action has been evaluated.
- Agent authority is scoped to the task rather than inherited from standing principal or workload privilege.
- Permitted actions distinguish materially different consequences rather than relying only on broad read/write permissions.
- Resource, data, tenant, and contextual scope are explicitly bounded.
- Transaction, rate, volume, duration, and consumption ceilings are defined where applicable.
- Higher-consequence stages can receive authority progressively rather than at task initiation.
- Persistent memory cannot silently preserve expired authority or approval.
- General-purpose tools do not create a larger effective authority envelope than the task requires.

CHAPTER 5

Tools, Data, and the Execution Boundary

An agent can think badly without changing anything.

It can misunderstand a request, retrieve an irrelevant document, choose an inefficient plan, or generate an entirely incorrect answer. Those failures may matter, but they remain fundamentally different from a failure that changes an account, sends a payment, modifies a repository, discloses protected information, or creates a purchase order.

The architectural transition occurs when generated intent crosses into protected enterprise state.

That crossing may look deceptively ordinary. The agent invokes an API. It calls a tool. It submits a query. It writes to a queue. It asks an MCP server to perform an operation. From the perspective of the surrounding infrastructure, these may be familiar integration patterns. Enterprise systems have been receiving API calls, database requests, messages, and commands for decades.

What changes is how the request was selected.

In conventional software, developers generally determine in advance which code path invokes which operation under which conditions. An agent can instead interpret a goal, select from available capabilities, construct parameters, incorporate retrieved information, and decide during execution which action to attempt. The enterprise therefore cannot treat the agent's decision to invoke a capability as evidence that the invocation should be permitted.

The tool boundary is where that distinction becomes consequential.

The agent may propose. The enterprise system acts.

That makes the execution boundary one of the most important architectural boundaries in an agentic system.

Tools Turn Reasoning into Consequence

The word *tool* can make an agent capability sound less consequential than it is.

A weather lookup is a tool. So is a function that reads a customer record. But a tool can also issue a refund, modify payroll data, create a purchase order, deploy software, reset an account, cancel a reservation, alter an access-control list, or initiate a payment.

Architecturally, these are not merely extensions of the model. They are interfaces to authority.

A useful definition is therefore broader than any particular agent framework: *A tool is a capability through which an agent can observe or affect state outside its own reasoning process.*

Some tools are informational. Others are transactional. Some wrap a single API operation. Others hide a substantial workflow. An MCP server may expose multiple tools and resources. A local function may ultimately invoke several enterprise services. A repository tool may expose read operations alongside operations that create branches, modify files, or initiate pull requests.

The abstraction presented to the agent does not determine the security significance of the capability.

The underlying consequence does.

Consider the procurement agent introduced earlier in the book. Jesse Fox's task requires it to inspect an approved request, identify the approved vendor, obtain the $3,850 catalog price, and prepare a purchase order for CC-4410. It might be given tools called:

```
search_catalog
```

get_supplier

check_budget

create_purchase_order

To the model, these may appear as four callable capabilities accompanied by descriptions and parameter schemas. To the enterprise, however, they cross very different boundaries.

Searching a product catalog exposes information.

Retrieving a supplier may expose commercial or financial data.

Checking a budget reveals protected organizational state.

Creating a purchase order creates a financial commitment.

The fact that all four appear in the same tool list does not make them equivalent.

This is why tool discovery and tool authorization must remain separate concepts. Discovery answers whether the agent knows that a capability exists and how to invoke it. Authorization answers whether this actor, performing this task, under this delegated authority, may use that capability in this way against this resource at this time.

Those questions must not collapse into one.

Availability Is Not Permission

Agent frameworks necessarily provide models with descriptions of available capabilities. The agent needs enough information to decide which tool is relevant and how to construct an invocation.

That description is part of the reasoning environment.

It is not a security boundary.

A statement such as: *"Use this tool only for purchases below $4,000."* may be useful guidance to the model. It may reduce undesirable tool selection. It may even be incorporated into evaluation and testing.

But it does not enforce a $4,000 purchasing limit.

If the underlying operation accepts a $5,200 purchase order and the caller possesses sufficient authority to create it, then the limit exists only as an instruction that the probabilistic actor is expected to obey.

The same problem appears with system prompts:

Never retrieve records belonging to another tenant.

Do not send customer information to external services.

Always ask the user before deleting a repository.

Do not modify production.

These are sensible behavioral instructions. They are not substitutes for tenant isolation, data authorization, egress controls, approval gates, or production access controls.

This distinction is particularly important because modern models can often follow instructions remarkably well. High reliability can create architectural complacency. A control that succeeds in 99.9 percent of ordinary tests is still not an enforcement boundary if the protected system will accept the prohibited operation when that remaining case occurs.

The governing question is not whether the agent *usually chooses correctly*.

It is what happens when it does not.

This is the same separation that has long existed between application intent and resource protection, but agents make the distinction harder to ignore. NIST SP 800-207 frames access around individual protected resources and least privilege rather than implicit trust derived from network position. The agentic extension of that principle is clear: neither the agent's location nor its own decision that an operation is appropriate should confer access to the resource.

The enterprise must protect the resource independently of the reasoning that selected it.

The Execution Boundary

The execution boundary is the point at which an agent's proposed action encounters a system capable of enforcing enterprise constraints before protected state is observed or changed.

It is not necessarily one physical component.

For one application, the boundary may be an API gateway followed by application authorization. For another, it may be an MCP server enforcing access before invoking an enterprise API. For a database operation, enforcement may occur in an intermediary service and again within the data platform. For a deployment agent, the important boundary may exist at the source-control platform, CI/CD system, production control plane, or all three.

The architecture matters more than the product.

On one side of the boundary is *agent-controlled selection:* reasoning, planning, tool choice, argument construction, and proposed action.

On the other side is *protected enterprise consequence:* data disclosure, state change, transaction creation, external communication, resource consumption, or some other effect for which the enterprise remains accountable.

Between them must be a place where the proposed operation can actually be prevented.

This yields a simple but demanding design rule:

A restriction on agent behavior is not an enforcement control unless the architecture contains a point outside the agent's reasoning where violation of that restriction can be stopped.

Later, we will examine the policy architecture behind those decisions. For now, the important issue is locating the boundary correctly.

If the only thing preventing an agent from issuing a $100,000 refund is a prompt telling it not to do so, there is no meaningful transaction boundary.

If the only thing separating one tenant's records from another's is context telling the model which tenant it is serving, there is no meaningful tenant boundary.

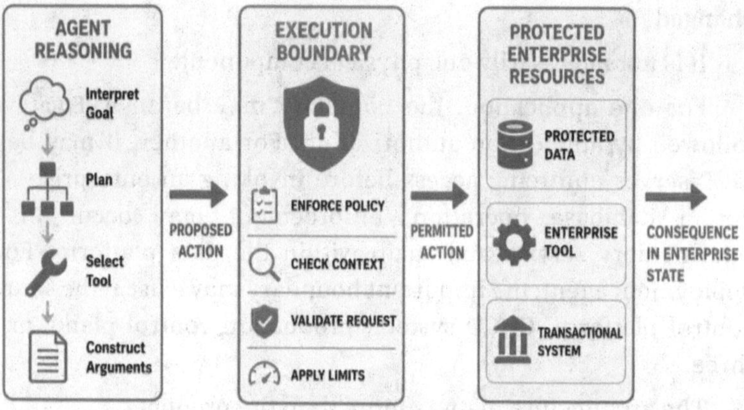

Agent Execution Boundary

The agent may select and propose an action, but protected enterprise state is reached only through an independently enforced boundary

If a development agent is told not to deploy directly to production but possesses credentials that allow it to do exactly that, there is no meaningful deployment boundary.

The system contains behavioral guidance, but the protected resource remains exposed to whatever request the agent eventually constructs.

Data Is Part of the Execution Surface

Tools are the obvious execution surface because their invocations are visible actions. Protected data creates a subtler version of the same problem.

An agent may require access to substantial information to complete a task. A customer-service agent may need account history. A financial agent may need balances and transaction records. A software agent may need source code, secrets metadata, logs, and deployment information.

The mistake is to assume that because information helps the agent reason, the agent should be allowed to retrieve all potentially useful information into its context.

Retrieval is an action.

If a principal would not be authorized to read a record directly, the agent acting for that principal should not acquire the record merely because the model believes it would be useful. Likewise, the fact that an agent has legitimate access to one body of protected information does not imply that all of that information may be combined with every other context available to the agent.

The relevant boundaries may include user, tenant, business unit, legal entity, geography, classification, purpose, session, and task.

A multitenant administrative agent makes the issue especially clear.

Suppose an operator asks an agent to investigate a configuration problem for Tenant A. The agent has a diagnostic tool capable of searching configuration records across the entire SaaS platform. The runtime may legitimately possess system-level access because the underlying service performs administrative operations.

But the agent's current task should not thereby acquire platform-wide authority.

The task context must constrain which data may cross into the execution. Tenant A is not merely a helpful parameter supplied to the model. It is an authorization boundary that must survive from task context into the protected data operation.

Otherwise, a hallucinated identifier, ambiguous query, poisoned retrieval result, or malformed tool argument can turn a legitimate diagnostic task into cross-tenant access.

This is one reason the identity work of the earlier chapters matters operationally. The protected system may need more than the runtime identity to make the correct decision. It may need the delegated principal, agent identity, task, tenant, purpose, or other authority-relevant context.

But that context must arrive as trustworthy security context, not simply as prose generated by the agent.

Context Is Not Automatically Trusted

Agentic systems consume information from many places.

User prompts, retrieved documents, web pages, databases, email messages, tool responses, memory stores, other agents, system instructions, and application state may all become part of the context from which the agent decides what to do next.

Traditional applications often make a strong distinction between data and executable instructions.

Language models make that distinction less dependable.

Text retrieved as *data* can influence subsequent behavior because the same model interpreting the task also interprets the retrieved content. A document can contain instructions. A tool response can contain instructions. An external source can attempt to redirect the agent toward another action.

This is the architectural significance of indirect prompt injection and tool poisoning. The problem is not simply that the model might produce a bad answer. Untrusted information entering the reasoning context can influence which privileged capability the agent invokes next.

OWASP describes MCP tool poisoning as a trust gap in which malicious or compromised tool content can introduce instructions into model context and influence later privileged actions. Its guidance consequently emphasizes enforcing restrictions at the tool execution layer rather than relying on system-prompt instructions to constrain privileged access.

The broader lesson applies regardless of protocol.

Context can inform an authorization decision, but untrusted context must not manufacture authority.

A purchase request may provide the amount the agent wants to spend. It does not establish the agent's spending limit.

A retrieved customer record may contain an account identifier. It does not establish permission to retrieve another account.

An email may instruct an agent to send a document elsewhere. It does not grant disclosure authority.

A tool result may recommend invoking another capability. It does not grant permission to invoke that capability.

Authority must come through the delegated and enforced architecture, not emerge from the text being interpreted.

A Compact Threat Model for Agentic Execution

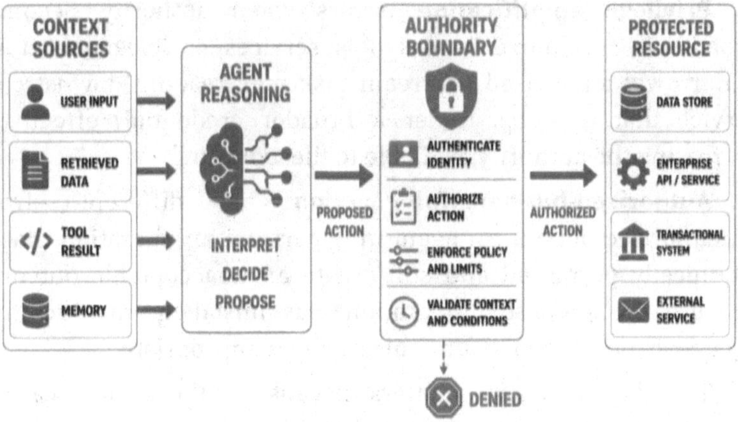

Context Influences What the Agent Proposes
Context Does NOT Grant Authority

Several recurring failure modes follow from this boundary problem. They will appear throughout the next few chapters, so it is useful to name them now.

Confused deputy behavior occurs when an agent or intermediary possesses authority that the requester does not and is induced to exercise that authority improperly. This is particularly dangerous when a broadly privileged runtime serves many users or tasks. OWASP's current agentic guidance explicitly identifies confused-deputy risk where an agent has higher privileges than the user and can be manipulated into exercising them.

Credential inheritance occurs when authority attached to the runtime, service, session, or initiating principal becomes available to the agent more broadly than the task requires. The agent does not need to defeat a control if the architecture has already handed it excessive authority.

Tool or context poisoning occurs when information presented as tool metadata, retrieved content, tool output, memory, or other context influences the agent toward an unsafe action. The dangerous step occurs when poisoned reasoning can reach a capability with sufficient authority to create consequence.

Privilege amplification occurs when authority becomes broader as work moves across tools, services, or delegated actors. A narrowly authorized upstream task may reach a downstream service that executes under a broader credential, effectively increasing the authority available to the workflow.

Authorized-but-harmful action is different from unauthorized action. An agent may perform an operation that is technically permitted and still create an unacceptable outcome because the requested operation was mistaken, manipulated, excessive, poorly timed, or contextually inappropriate.

This last category matters because authorization cannot answer every question about safe behavior.

Suppose an agent is legitimately permitted to issue customer refunds up to $500. It refunds $475 to the wrong customer because it misinterprets the conversation.

The action may have been authorized.

It was still harmful.

Or suppose an agent is permitted to email an approved report to a customer. It sends the correct report to the wrong authorized contact.

Again, access control alone may not detect the semantic mistake.

This is why agent security cannot be reduced to stronger IAM. Identity and authorization are necessary foundations, and NIST's current agent-identity work is explicitly examining identification, authorization, auditing, and prompt-injection mitigation for agents. That work remains an evolving effort rather than a finalized agent-security standard.

The architectural objective is therefore not to pretend that deterministic authorization can determine whether every agent decision is wise. It is to ensure that errors, manipulation, and ambiguity cannot freely translate into unlimited consequence.

OWASP's excessive-agency guidance makes essentially the same risk visible from another direction: damaging actions can result from unexpected or manipulated model behavior when the system grants the model capabilities and permissions beyond what is necessary.

That is precisely why the authority envelope established earlier must meet an execution boundary here.

Provenance Matters When Data Influences Action

Not every piece of context deserves equal trust.

An account balance obtained directly from the system of record is different from a balance copied into an email. A supplier bank account retrieved from an approved master-data system is different from one appearing in an invoice attachment. A deployment instruction originating from an approved change record is different from one found in repository documentation.

To the model, all may appear as text or structured data.

To the architecture, provenance can be material.

This does not require a universal provenance infrastructure for every token entering a model. It does require architects to identify where the origin and integrity of information affect the authority or consequence of an action.

A procurement agent might be permitted to use an invoice to determine an invoice number and amount while being prohibited from treating payment instructions contained in that invoice as authoritative banking information.

A development agent might use repository documentation to understand how an application works while being prohibited from treating instructions in source comments as authorization to retrieve production secrets.

A customer-service agent might summarize user-supplied information without allowing that information to change the identity of the customer whose protected records it can access.

The architectural question is therefore not only what information the agent can retrieve. It is what role that information is allowed to play.

Data used as evidence for a consequential action may require stronger provenance than data used merely to help construct a response.

The Boundary Belongs Near the Consequence

It is tempting to centralize all agent controls around the orchestration layer.

That layer has excellent visibility. It knows the task, model, tools, context, and current plan. It can filter available tools, validate schemas, inspect requests, and enforce useful constraints.

Those controls are valuable. But they should not become the only enforcement point for consequential operations. The farther enforcement is from the protected resource, the more opportunities exist for assumptions to diverge.

A tool gateway may believe the current tenant is Tenant A while a downstream API accepts an arbitrary tenant identifier. An orchestrator may limit purchase orders to $4,000 while a downstream procurement API accepts any amount available to the credential. An agent platform may prohibit production deployment while the source-control credential still permits modification of a workflow that deploys automatically.

If the answer is another instruction to the model, the execution boundary is incomplete.

If the answer is that the agent "should never do that," the execution boundary is incomplete.

If the answer is that the tool was not intended to be used that way, the execution boundary is incomplete.

The architecture becomes meaningful when the answer identifies an independently enforceable constraint at or near the protected resource.

This is not an argument that every action should be blocked by a complicated centralized policy engine. Many actions can be controlled using familiar mechanisms: scoped API authorization, tenant-aware services, transaction constraints, database permissions, workflow states, capability restrictions, approval requirements, quotas, schemas, or application-level invariants.

The particular mechanism depends on the resource.

The architectural requirement is that the control exists outside the agent's own reasoning.

MCP at the Execution Boundary

The Model Context Protocol is one contemporary way to expose tools and resources to agentic applications. Architecturally, its importance is not that it creates a new kind of enterprise authority. It places a recognizable client/server boundary around capabilities that an agent can discover and invoke.

An MCP server should therefore be treated as a protected capability surface. A tool appearing in a server's catalog means that the capability can be described to the client; it does not establish that every connected agent may use it. Discovery, authentication, authorization, invocation, and downstream transaction enforcement remain separate concerns.

Where OAuth-based authorization is used, the authorization server can issue access appropriate to the client and protected resource. The enterprise still must decide whose authority is represented, how the agent and task are bound to that authority, which tools and resources are permitted, and whether the target system must perform an additional transaction-level decision. A valid access token is not proof that a particular purchase, deployment, disclosure, or account change is authorized.

Enterprise-managed authorization may also impose policy outside the MCP server: at a gateway, tool broker, policy enforcement point, target API, or system of record. This is particularly important when one server fronts capabilities with very different consequences. The server boundary can concentrate control, but it should not become a giant undifferentiated permission.

MCP specifications and authorization guidance continue to evolve. Implementations should record the protocol and server versions in use and avoid treating current mechanism details as timeless architecture. The durable principle is that protocol connectivity never substitutes for task-scoped authority and enforcement at the consequence boundary.

Architect's Takeaway

Agent architectures should not be reviewed primarily by examining what the agent has been instructed to do. They should be reviewed by examining what the surrounding systems will actually allow it to do. That changes the architecture conversation.

A tool catalog is not merely a list of useful capabilities. It is a map of potential execution surfaces. Retrieval is not merely grounding. It is protected data access. Persistent memory is not merely convenience. It may carry authority-relevant state across task boundaries. Tool results and retrieved documents are not merely context. They may be untrusted inputs capable of influencing subsequent privileged actions.

Most importantly, the place where the model chooses an action and the place where the enterprise authorizes that action should be treated as different architectural concerns.

The agent can decide what it wants to do.

The enterprise still needs somewhere to say no.

Authority Check

- Where, specifically, does agent-generated intent cross into protected enterprise state?
- Can a discovered or described tool be invoked even when the current task should not authorize it?
- Are tenant, user, task, and data boundaries enforced outside the model's context?
- Can untrusted retrieved content or tool output influence a more privileged downstream action?
- Can persistent memory carry approvals, permissions, tenant context, or other authority-relevant state beyond the task that established them?
- Which systems ultimately control the consequential state, and can they independently reject an invalid operation?

Decision Checklist

- Every consequential tool invocation crosses an identifiable enforcement boundary.
- Tool discovery is treated separately from permission to invoke the tool.
- Prompts and tool descriptions are not relied upon as security controls.
- Protected data access is constrained by enforceable user, tenant, task, or other applicable boundaries.
- Untrusted context cannot independently create or expand authority.
- Persistent authority-relevant state has explicit lifetime, provenance, and scope.
- Transactional systems enforce constraints at or near the system that owns the consequence.
- The architecture remains bounded when the agent proposes an incorrect but syntactically valid action.

CHAPTER 6

Deterministic Control and Human Authority

Earlier chapters have established the execution boundary: the point at which an agent's generated intent becomes an attempt to access protected data, invoke a capability, or change enterprise state. At that boundary, the enterprise needs more than a control point. It needs an authorization decision.

The difficult part is determining what that decision should represent.

For an ordinary API request, authorization may be expressed largely in terms of a subject, resource, and operation. Agentic execution can require a richer decision. The actor making the call may be an agent operating under authority delegated by another principal, for a particular task, within limits that apply only to this execution. The requested action may also carry transaction-specific consequences that cannot be determined from the API operation alone.

The authorization question therefore becomes: *May this agent, acting under this delegated authority, perform this action against this resource for this task under the current conditions?*

That is the problem deterministic enforcement must solve.

The Agentic Authorization Request

Consider the procurement agent from earlier chapters. The agent proposes Jesse Fox's purchase order: one approved laptop, $3,850, cost center CC-4410, ship to Portland.

Knowing that the caller is the procurement agent does not equate to ability to authorize the transaction. The same agent might legitimately create that purchase order for one principal but not another, against one cost center but not another, or as part of one delegated task but not another.

A useful authorization request might therefore contain:

- originating principal;
- agent identity;
- execution or instance identity where relevant;
- delegation or authority context;
- task;
- requested operation;
- target resource;
- transaction attributes;
- environmental or operational conditions.

Not every authorization decision requires every element. A read operation against low-risk reference data may require considerably less context than a financial transaction. The point is not to construct the largest possible policy request. It is to provide the information necessary to determine whether the specific exercise of delegated authority is permitted.

This creates an important connection to the architectural chain developed throughout the book:

Principal → Delegation → Agent Identity → Task → Policy → Tool → Action

At execution time, portions of that chain become inputs to authorization.

That makes preserving the chain operationally useful, not merely valuable for later audit.

Policy Decision and Enforcement

Two established architectural roles are particularly useful here.

The *Policy Decision Point*, or PDP, evaluates the authorization request against policy.

The *Policy Enforcement Point*, or PEP, applies the resulting decision where the action can actually be permitted or prevented.

These roles need not correspond to separate products or services. A service may contain its own enforcement point while consulting a local or remote policy engine. An API gateway may enforce coarse-grained resource access while the protected application performs a more transaction-specific decision. A distributed policy engine may evaluate centrally managed policy close to the resource.

What matters is the separation of responsibility:

The PDP determines whether the proposed action is permitted. The PEP ensures that the decision governs execution. The PEP must therefore exist on a path the agent cannot bypass.

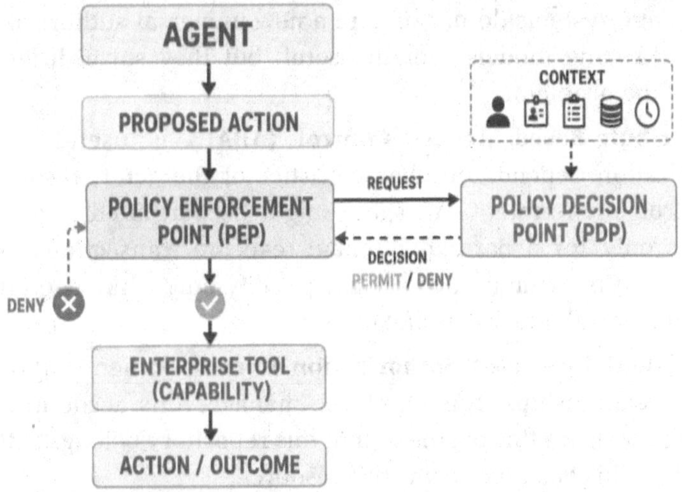

Policy Decision and Enforcement Points

The agent selects a proposed action, but an independent policy decision determines whether that action may cross the enforcement boundary and reach the protected enterprise capability.

For the procurement example, an upstream gateway might establish that the agent is permitted to reach the purchasing service. That does not necessarily establish that it may create this purchase order. A transaction-level enforcement point may still need to evaluate the supplier, amount, business unit, cost center, approval state, and delegated authority immediately before the order is committed.

Authorization becomes more precise as it approaches consequence.

That does not imply that every system needs layers of policy engines. The appropriate number of enforcement points follows the resource and transaction boundaries. The architectural requirement is simply that policy be enforced at the point where the prohibited consequence can still be prevented.

Choosing the Authorization Model

Agentic systems do not require a new universal authorization model. Existing models remain useful, but they solve different parts of the problem.

Attribute-Based Access Control (ABAC) is useful when authorization depends on characteristics of the actor, resource, action, or environment. An agent might be permitted to access records only for a particular tenant, execute transactions only within a defined amount, or use a capability only while operating under a particular task classification.

Relationship-based authorization is useful when authority follows relationships: this employee manages this account, this agent operates for this business unit, this repository belongs to this project, or this customer owns this resource.

Capability-oriented approaches are useful when authority is deliberately conveyed as a constrained ability to perform particular actions. This maps naturally to task-bound authority because the authority carried into execution can itself be narrow.

Transaction authorization becomes important when permission depends upon the actual operation being committed.

The question is no longer simply whether the agent may call CreatePurchaseOrder, but whether it may create *this* purchase order with *these* consequential attributes.

An enterprise architecture may use several of these simultaneously. Relationship information may establish which resources are eligible, attributes may constrain the circumstances of access, delegated credentials may carry limited authority, and transaction rules may determine whether the final operation can proceed.

The design question is therefore not, *Which authorization model should we standardize on for agents?*

It is: *Which facts determine whether this particular delegated action is authorized, and which policy model expresses those facts cleanly?*

Policy Inputs Must Be Trustworthy

A deterministic policy engine can still produce the wrong security result if its inputs are untrustworthy.

Suppose policy permits Jesse's $3,850 order to execute without further approval, and requires a manager approval for any purchase above $4,000. The rule is deterministic. But where does the approval state come from?

If the agent can simply include `managerApproved = true` then the policy has not established meaningful control.

The same problem applies to tenant identifiers, task classifications, delegation scope, transaction limits, risk levels, principal identity, and other authority-relevant attributes.

Some context originates with the agent because the agent is proposing the action. That is unavoidable. The proposed purchase amount, for example, may originate in the tool invocation the agent constructed.

Other context should come from authoritative systems.

Principal identity may come from the identity and delegation infrastructure. Task identity may be established by the orchestration system. Approval state may come from a workflow service. Tenant relationships may come from enterprise data. Remaining transaction authority may come from a policy or transaction system rather than from the agent's memory.

This produces a simple rule:

The more important an attribute is to the authorization decision, the less willing the architecture should be to accept the agent's unsupported assertion of it.

This is especially important because agents operate on context. Prompt content, retrieved documents, tool responses, memory, and other inputs can influence what the agent believes to be true. A poisoned document may convince an agent that a transaction was approved. That should not make the approval true from the authorization system's perspective.

The PDP should evaluate authoritative facts, not merely the agent's interpretation of those facts.

Contextual Authorization

Agent authority can also depend on conditions that change during execution.

A task may begin at 9:00 a.m. and continue for hours. During that period:

- the principal's access may change;
- the agent may consume part of its transaction allowance;
- an approval may expire;
- a resource may change classification;
- a risk condition may change;
- the agent or task may be suspended.

An authorization decision made when the task begins cannot always authorize everything that happens later.

The appropriate response is not continuous policy evaluation of every reasoning step. Internal reasoning is not normally the protected consequence. Instead, reevaluation should occur where authority is exercised and where changed conditions matter.

A long-running agent may therefore authenticate once, receive task-bound authority, and still encounter fresh authorization decisions when it reads sensitive information, initiates transactions, changes protected state, or crosses another significant boundary.

Asynchronous execution deserves particular attention. If an agent schedules an operation for later execution, the enterprise must decide whether authorization is established when the work is queued, when it executes, or both.

For low-consequence work, authorization at submission may be sufficient. For consequential transactions, conditions may need to be reevaluated immediately before execution.

The governing question is whether stale authorization can create an unacceptable consequence.

Denial Is Part of the Architecture

Adaptive agents introduce another consideration that deterministic applications rarely present in the same way: the agent can respond creatively to failure.

Suppose an agent attempts an operation and the PEP denies it. A planning agent may interpret that result as an obstacle and search for another way to accomplish the task.

That behavior can be useful when the failure is operational. If one information source is unavailable, using another may be exactly what the enterprise wants.

Authorization denial is different.

If the agent is forbidden to perform an action, reaching the same consequence through another tool or API is not successful adaptation. It is authorization bypass.

The architecture therefore needs consistent protection across equivalent paths. If creating a purchase order requires a particular authority, that requirement should apply whether the operation arrives through an ERP API, an MCP-exposed tool, an integration service, or another reachable interface.

The agent also needs defined behavior when authority is insufficient. Depending on the task, it may:

- select a genuinely different action that remains within its authority;
- reduce the scope of the proposed operation;
- request additional authority through an established mechanism;
- escalate for approval;
- terminate the affected portion of the task.

What it should not do is infer that a different technical route changes the authorization boundary.

This distinction becomes increasingly important as agents gain better planning and tool-discovery capabilities. The more capable the agent becomes at finding ways to accomplish a goal, the more consistently the enterprise must enforce what it is actually allowed to accomplish.

Policy as an Enterprise Control Plane

Authorization logic can be embedded separately into every tool and application, but agentic systems make the limitations of that approach increasingly visible.

An agent may interact with APIs, databases, MCP servers, SaaS applications, repositories, messaging systems, and transactional platforms during a single task. If each system interprets delegated authority differently, the enterprise cannot easily establish consistent boundaries.

Policy-as-code and external policy decision services provide one architectural response. They allow authorization rules to be managed independently of agent reasoning and, where appropriate, consistently applied across multiple enforcement points.

This does not mean every authorization decision must traverse one centralized service. That architecture could introduce unacceptable latency and availability dependencies. Policy may be distributed to local decision points, cached where appropriate, or implemented through resource-native controls.

The important centralization is governance of policy semantics, not necessarily execution of every policy decision.

The enterprise should be able to answer questions such as:

- Which policy governs this class of agent action?
- Who owns that policy?
- Which attributes does it depend upon?
- Which systems enforce it?

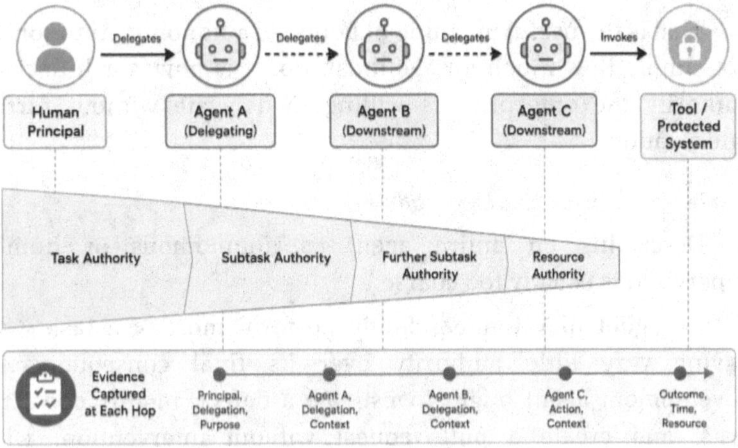

Agent-to-Agent Delegation and Authority Attenuation

Authority narrows as work is delegated downstream; each agent receives only the authority required for its assigned portion of the task.

- How are policy changes versioned and deployed?
- Can the policy decision associated with a consequential action later be reconstructed?

Those questions become increasingly important as authority crosses agents and systems.

Human Authority and Bounded Autonomy

By this point in the architecture, the agent has an established identity, legitimately delegated authority, defined boundaries, and independent authorization around consequential actions. None of that answers one remaining question: *Which decisions should the agent be allowed to make alone?*

An agent may be authorized to access a procurement system, prepare a purchase order, and even possess the technical capability to submit it. The enterprise must still decide whether submission is within the agent's independent decision authority or requires another actor to authorize the consequence.

That is the useful meaning of bounded autonomy. Autonomy is not simply how much an agent can do. It is how much decision authority the enterprise is willing to delegate without further intervention.

Autonomy Is Delegated Decision Authority

Describing an entire agent as autonomous or human-supervised is usually too coarse.

An agent may independently perform most of a task while having very little authority over its final consequence. A development agent might investigate a defect, modify code, run tests, and create a pull request without intervention, while remaining unable to decide that its changes should enter production. A procurement agent might research suppliers and prepare an order while lacking authority to commit the organization to the purchase.

Conversely, requiring occasional human interaction does not necessarily make an agent meaningfully supervised. A human may approve a broad objective at the beginning of a task while the agent subsequently makes dozens of consequential decisions without further intervention.

The relevant boundary is therefore not between "autonomous agent" and "human-in-the-loop agent." It is between decisions the enterprise has delegated to the agent and decisions it has retained for another authority.

That boundary should follow consequence.

Low-consequence decisions can often be delegated broadly. As potential impact increases, the case for retaining human decision authority becomes stronger. The threshold will vary by enterprise and by transaction. A $50 customer adjustment, a $50,000 purchase, and a change to production security policy should not receive identical treatment simply because the same agent can technically perform all three.

Model capability does not determine that threshold. Neither does model confidence. A highly capable agent can still reach a decision that the organization has deliberately reserved for a human.

The autonomy boundary is therefore an enterprise authority decision, not a measure of AI sophistication.

Put the Human at the Consequence Boundary

Human approval is most useful where the agent reaches the end of the decision authority already delegated to it.

Suppose a procurement agent has independently assembled a valid purchase order. It has selected an approved supplier, verified the price, confirmed the accounting information, and satisfied the applicable policies. If the purchase exceeds the authority the enterprise has delegated to the agent, the question presented to the human is not whether the agent performed its previous work correctly.

The question is whether the organization should commit to the purchase.

That is a genuine authority transition.

The same pattern appears elsewhere. A development agent can prepare a deployment, but production release may remain a human decision. A customer-service agent can calculate a proposed account adjustment, but an unusually large credit may require a supervisor. An administrative agent can identify accounts that appear eligible for deletion, while final deletion remains reserved for another authority.

This placement matters because human involvement inserted earlier or later may add little control. Asking a person to approve routine intermediate steps wastes attention. Asking for approval after the consequential decision has effectively been made turns oversight into ceremony.

The strongest approval point is usually where the transaction crosses from preparation into commitment.

Not every workflow has one clean boundary. Some contain several consequences of different magnitude. The architecture may therefore contain more than one autonomy threshold. What matters is that each threshold represents a deliberate decision about who is authorized to decide what happens next.

Reversibility Changes the Decision

Consequence is partly a function of reversibility.

An incorrect draft can be discarded. A database change may be restored. A deployment may be rolled back. Money sent outside the enterprise, confidential information disclosed to the wrong party, or communication delivered to thousands of customers may be much harder to recover.

That difference should influence the autonomy boundary.

Where practical, an architecture can allow the agent substantial autonomy on the reversible side of a consequential transition while retaining additional authority at the point of commitment. This preserves much of the economic value of autonomous work without requiring the enterprise to delegate every final decision.

Reversibility is not a guarantee of safety. Rollback may be incomplete, expensive, or itself consequential. But it is a useful factor when deciding how much independent decision authority an agent should receive.

The question for the architect is not simply, *Can this action be undone?*

It is: *What would it cost the enterprise if this decision were wrong, and how reliably could the resulting consequence be recovered?*

When Human Oversight Becomes Theater

Adding an approval step does not automatically create meaningful human oversight. The project's governing distinctions explicitly reject the assumption that human approval makes an unsafe action safe.

A human cannot meaningfully authorize a decision that is not understandable from the information presented. An approval request that says only *Agent requests permission to continue* provides little basis for judgment. The approver needs to understand the consequence being authorized.

Volume creates another problem. If agents generate hundreds of routine approval requests, humans adapt. Requests that are almost always safe become requests that are almost always approved. The organization retains an impressive approval history while losing much of the judgment the approval was intended to provide.

This is approval fatigue, and it is partly an architecture problem.

The remedy is not to place humans everywhere. It is to give agents appropriate authority for routine, acceptable consequences and reserve human attention for decisions where additional judgment or authority is actually required.

Meaningful oversight therefore has several characteristics. The human understands the proposed consequence, has sufficient information to make the decision, possesses the authority being requested, and can actually prevent the action from proceeding.

Without those properties, the human may be in the workflow without meaningfully being in control of it.

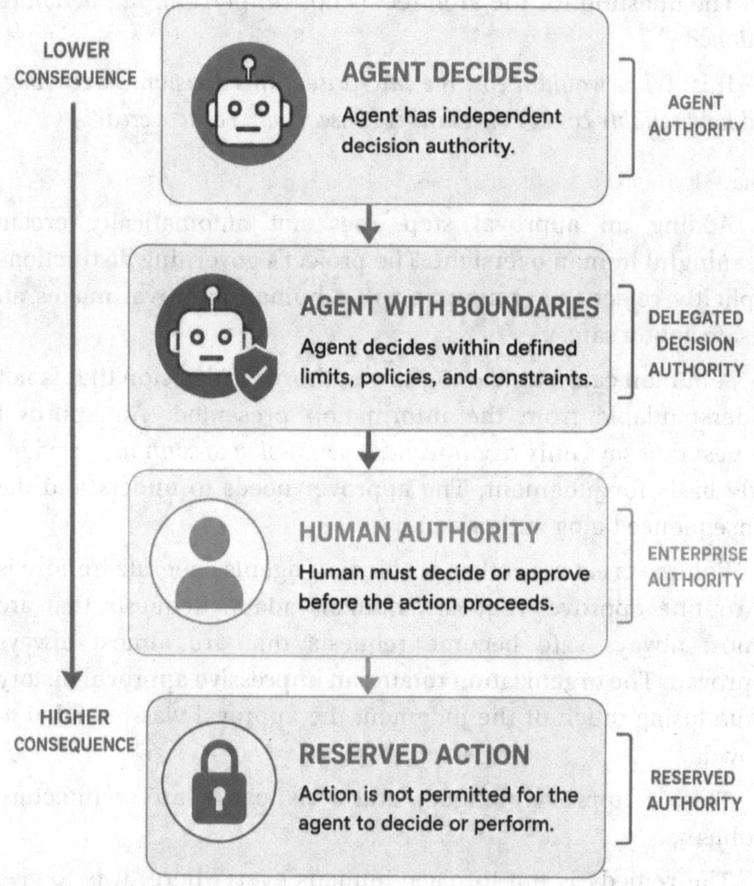

Bounded Autonomy

Agents can often operate with greater autonomy while an action remains preparatory or readily reversible; stronger authority may be appropriate when the action crosses into consequential commitment.

Preserve Separation of Duties

Agent automation can also collapse authority boundaries that the enterprise intentionally created.

A process may require one party to initiate a transaction and another to approve it. Replacing both activities with a single agent defeats that separation even if the agent performs each step correctly.

Drawing two agents instead of one does not necessarily solve the problem. If both operate under the same effective authority and one automatically accepts the other's decisions, the architecture may have created two software components without creating independent authority.

The relevant question is whether the approval represents an independent decision by an appropriately separate authority.

This is particularly important when agent adoption begins by automating existing human workflows. The architecture team should identify which human steps exist merely because humans historically performed the work and which exist because the organization deliberately separated authority.

The former may be candidates for elimination.

The latter are security or governance boundaries and should not disappear accidentally.

Choosing the Autonomy Boundary

The appropriate autonomy boundary depends less on how impressive the agent is than on what happens when its decision is wrong.

A simple decision matrix can make that explicit:

Potential consequence	Recoverability	Appropriate decision authority
Low	Easy	Agent
Moderate	Easy	Agent, with defined boundaries
Moderate	Difficult	Agent or human depending on context

Potential consequence	Recoverability	Appropriate decision authority
High	Reliable recovery available	Additional authority usually appropriate
High	Difficult or irreversible	Human or separately accountable authority

This is not a universal policy. A financial institution, healthcare organization, manufacturer, SaaS provider, and internal development platform will draw different lines.

The matrix instead establishes the architectural method.

Start with the consequence. Consider its magnitude, scope, reversibility, and recoverability. Then decide whose judgment is required.

This avoids a common inversion in agent programs: deciding first that the organization wants "more autonomy" and then searching for controls that make the desired autonomy acceptable.

The better direction is the reverse.

Determine the consequence the enterprise is prepared to delegate, then grant the autonomy appropriate to that consequence.

Architect's Takeaway

Human involvement should not be scattered through an agent workflow as evidence of responsible design. It should appear where human authority is actually required.

The architecture review question is therefore not *"Is there a human in the loop?"* It is *"Which decisions has the enterprise delegated to the agent, and where does that delegated decision authority end?"*

A well-designed agent may perform substantial work without human involvement while stopping at a small number of carefully chosen consequence boundaries. That is not less governed than a workflow filled with approval prompts. It is usually more explicit about where authority actually resides.

Authority Check

- Which consequential decisions can the agent make?
- Where does preparation become commitment?
- Which consequences justify retaining human decision authority?
- How reversible and recoverable are autonomous actions when they are wrong?
- Does the approver have enough information and actual authority to make the requested decision?
- Has automation accidentally eliminated an existing separation-of-duties boundary?

Decision Checklist

- Autonomous decision authority is explicitly defined.
- Human intervention occurs at meaningful consequence boundaries.
- Reversibility and recoverability influence the autonomy decision.
- Routine approvals have been removed where they add no meaningful judgment.
- Approval requests provide sufficient information to understand the consequence.
- Required separation of duties survives automation.
- Autonomy follows acceptable consequence rather than desired AI capability.

chosen consequence boundaries. This is not less control [illegible] [illegible] with approval [illegible], [illegible] operation [illegible] actually ready.

- Which consequences determine [illegible] agents [illegible]

- Where control parameters become commitments?
- Which consequences justify retaining human decision authority?
- How reversible and recoverable are autonomous actions [illegible]
- [illegible] authority [illegible]
- Has [illegible]

- [illegible] autonomous [illegible] autonomy [illegible]

- Human [illegible] decisions [illegible]
- Reversibility and recoverability influence the autonomy decision.
- Human approvals have been removed where [illegible] meaning [illegible]
- Approval [illegible] understand the consequences.
- Distributed supervision [illegible] automation.
- Autonomy [illegible] rather than [illegible] AI capability.

CHAPTER 7

Interruption, Revocation, and Containment

Preventing an agent from taking an unauthorized action is one architectural problem. Stopping an agent after legitimate authority has already been granted is another.

That distinction becomes important as agents move from short, synchronous interactions to longer-running work. An agent may hold credentials, maintain a session, submit asynchronous jobs, initiate workflows, create messages on queues, delegate work downstream, or cause transactions that continue after the original invocation has ended. At that point, denying the next API request may be necessary, but it may not be sufficient.

The architectural question changes from *Should this action be allowed?* to *How do we stop authority that is already in motion?*

An enterprise that delegates consequential authority to agents therefore needs more than authorization. It needs an interruption architecture.

The objective is not to guarantee that every consequence can be undone. Some cannot. The objective is to ensure that delegated authority can be withdrawn, active execution can be interrupted, pending work can be contained, and completed consequences can be identified for remediation.

This is what makes revocability an architectural property rather than merely a credential feature.

Authorization Has a Lifecycle

Earlier chapters established that an agent should receive bounded authority appropriate to its task. That authority may be constrained by resource, action, time, transaction value, context, and other conditions. Those controls determine what the agent is permitted to do.

But permission is not necessarily static for the lifetime of a task.

A user may cancel a request. An administrator may suspend an agent. A security system may detect suspicious behavior. A downstream resource may become compromised. A transaction may exceed an emerging risk threshold. An incident responder may decide that all activity associated with a particular agent version must stop immediately.

In each case, authority that was legitimate when issued is no longer legitimate now.

That creates a lifecycle:

grant → use → reevaluate → revoke or expire

Expiration handles one part of the problem. Short-lived credentials reduce the duration for which authority remains usable, but waiting for expiration is not the same as revocation. If a credential remains valid for another ten minutes and the enterprise needs the agent stopped now, ten minutes may represent hundreds of requests or several irreversible transactions.

More importantly, the credential is only one place where the agent's authority may have been converted into executable work.

The agent may already have submitted a job.

The job may already be on a queue.

A downstream agent may already be working.

An external system may already be processing a transaction.

Revoking the original credential does not necessarily recall any of them.

Revocation and Interruption Are Different Controls

It is useful to separate revocation from interruption.

Revocation changes whether authority remains valid. Interruption changes whether execution continues.

They frequently work together, but they solve different problems.

Suppose the procurement agent is submitting Jesse Fox's $3,850 laptop order under legitimately delegated authority. During execution, unusual activity is detected and an operator suspends the agent.

The enterprise might revoke or invalidate the credentials associated with the active task. That prevents those credentials from authorizing additional protected operations, assuming the relevant enforcement points recognize the change quickly enough.

But the agent process may still be running.

It may continue reasoning, generating requests, modifying local state, or producing work for later execution. Jobs already accepted by another system may continue independently. Work delegated to another agent may continue under credentials that were separately issued.

Stopping the runtime without revoking its authority creates the opposite problem. The current process may disappear, but usable credentials or resumable work may remain. Another instance might restart the task and continue.

Effective interruption therefore requires control over both authority and execution.

This leads to a simple architectural principle:

Stopping an agent means stopping its ability to continue producing consequential effects, not merely stopping one process or invalidating one credential.

There Is More Than One Thing to Stop

A production agent system usually contains several independently interruptible elements.

Credentials determine whether protected operations can continue to be authorized.

Tasks represent the unit of delegated work and may need to be cancelled regardless of which runtime currently owns them.

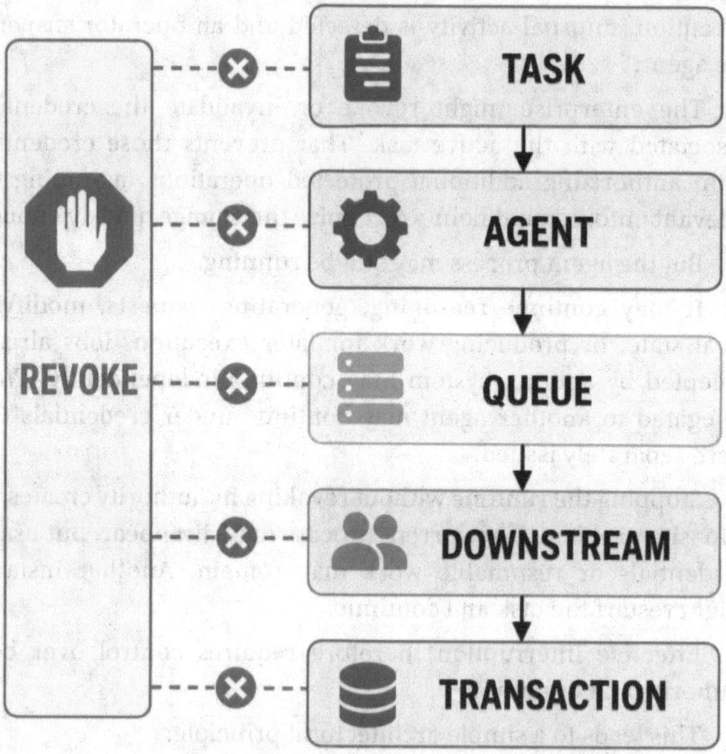

Interruption and Containment Paths

Revoking authority or terminating an instance may stop future execution at one boundary while queued work, downstream delegation, or already-committed transactions continue elsewhere.

Sessions may preserve authenticated or application-level state that allows execution to continue.

Agent instances may need to be terminated or isolated.

Queued work may already have crossed from the agent into asynchronous infrastructure.

Downstream delegations may have created independently executing tasks.

Transactions may already have reached systems whose state changes cannot simply be cancelled.

These controls should not be treated as interchangeable.

Terminating an agent instance does not necessarily terminate its task. Revoking a credential does not necessarily remove a queued command. Cancelling a parent task does not necessarily cancel work already delegated to a child. Suspending a logical agent does not automatically reverse transactions completed before the suspension.

The architecture must know which of these states exist and which interruption mechanism applies to each.

This is particularly important in distributed systems because there may be no single instantaneous "stop" operation. Interruption is usually a coordinated state transition across several boundaries.

The Task Is an Important Containment Unit

A task-bound authority model provides a useful basis for interruption because the enterprise can identify what work belongs to the authority being withdrawn.

Consider a task that instructs an agent to reconcile a set of invoices. During execution, the agent has:

- retrieved invoice records;
- requested supporting documents;
- delegated anomaly analysis to another agent;
- queued three corrections;
- completed one adjustment; and
- prepared two more transactions.

If the task is cancelled, the enterprise needs more than an instruction sent to the reasoning loop.

The cancellation should become authoritative state.

New actions associated with the cancelled task should be rejected. Active workers should detect the cancellation. Pending jobs should be identified. Downstream tasks should be evaluated for cancellation. Transactions that have not crossed their commit boundary should be stopped where possible.

This makes the task identifier operationally significant. It can serve as a correlation point across credentials, execution, downstream work, and evidence.

Without that continuity, containment becomes forensic guesswork: operators know that an agent must be stopped but cannot reliably determine which work belongs to the affected execution.

Revocation Must Reach the Enforcement Point

A revocation mechanism is useful only if the system capable of accepting the next action can recognize that authority has changed.

This is one reason short-lived credentials remain valuable. They place a natural upper bound on stale authority. But systems with consequential agent activity may require faster revocation semantics than expiration alone provides.

The implementation choices vary. A system might consult centralized authorization state, introspect credentials, maintain deny or suspension state, use session controls, or combine several mechanisms. The architecture does not require one universal technique.

It does require a defined answer to three questions:

What event withdraws the authority?

Where is that withdrawal recorded?

How quickly will every relevant enforcement point honor it?

The third question is often the difficult one.

A control plane may record an agent as suspended immediately while a resource server continues accepting a previously issued credential. A policy service may reject new operations while an already established session remains usable. A parent agent may know that a task was cancelled while a downstream agent remains unaware.

The interval between revocation and effective enforcement is a period of residual authority.

Architects should treat that interval as a measurable risk characteristic, not an implementation detail.

Asynchronous Work Changes the Problem

Synchronous execution creates an intuitive interruption model: stop the caller or reject the call.

Asynchronous systems are less cooperative.

An agent may publish a command and receive confirmation that it was accepted. The consumer may process it seconds or minutes later. Once accepted, that command has become an independently progressing unit of work.

A queue may therefore contain commands whose originating authority has since been revoked.

This means authorization may need to be evaluated not only when work is submitted but also when consequential work is executed.

Whether that is practical depends on the system. Some messages represent immutable events that should never be withdrawn. Others represent commands whose legitimacy depends on still-current authority. Those are architecturally different things and should not be modeled identically.

For agent-generated commands, useful execution context can include the originating principal, task, agent identity, authority reference, and relevant expiration or cancellation state. The consumer can then determine whether the work remains executable rather than assuming that acceptance into the queue permanently legitimized it.

This matters especially when the delay between submission and execution is significant.

Authority can change while a message waits.

Downstream Delegation Creates a Revocation Tree

Delegation to another agent adds another complication.

If Agent A delegates a portion of its task to Agent B, cancelling Agent A's current process should not be assumed to cancel Agent B. The downstream agent may have received separately issued, attenuated authority and may be running in another environment.

The enterprise therefore needs explicit semantics for downstream cancellation.

Those semantics need not always mean "cancel everything." Some downstream work may be safe or desirable to complete. A research task that has no consequential authority might continue even though a purchasing task is stopped. Conversely, an execution task capable of changing financial state may need immediate cancellation when its parent authority disappears.

The important architectural property is that the relationship is known.

A downstream delegation should retain enough linkage to determine where its authority originated, which parent task created it, and whether revocation of the parent should affect the child.

This turns delegation chains into revocation paths as well as authority paths.

Authority may attenuate as it moves downstream, but cancellation information often needs to propagate in the same direction.

Kill Switches Need Defined Scope

"Kill switch" is a useful phrase but an imprecise architecture.

What exactly does the switch kill?

A single task?

An agent instance?

Every instance of a logical agent?

All agents using a particular tool?

All activity associated with a compromised credential issuer?

Every autonomous transaction in an environment?

These are very different controls.

A useful containment architecture provides multiple scopes of interruption rather than one giant emergency stop.

A **task kill** stops a particular unit of work.

An **instance kill** terminates a particular execution.

An **agent suspension** prevents a logical agent from initiating or continuing protected activity.

A **capability suspension** removes access to a particular tool or resource.

A broader **environment or platform stop** may be appropriate for severe incidents.

The broader the switch, the larger its operational consequence. A global shutdown may be appropriate when the alternative is uncontrolled financial or security impact, but it should not be the only available containment mechanism.

Granular interruption allows the enterprise to stop the smallest affected authority domain first and expand containment when necessary.

Stopping Future Action Is Not Reversing Past Action

Interruption has a hard boundary: completed consequences.

If an agent has already sent an email, disclosed data, merged code, transferred funds, submitted an order, or triggered an external process, terminating the agent does not erase what occurred.

The architecture therefore needs to distinguish three states:

Preventable – the action has not yet crossed the enforcement boundary and can still be denied.

Interruptible – execution has begun, but the operation can still be cancelled before the consequential commit.

Completed – the consequence has occurred and requires remediation, reversal, or acceptance.

The last category is where compensating transactions become important.

Many enterprise systems do not support true rollback across distributed boundaries. Instead, they support a second action that compensates for the first: cancel the order, issue the refund, restore the prior configuration, revert the commit, disable the account, or create an adjusting transaction.

Compensation is not equivalent to erasure. The original action still occurred and should remain part of the evidence chain.

There are operational and legal implications to consider here. A refund does not mean a payment never happened. A revert does not mean code was never deployed. Revoking access does not mean information was never disclosed.

Containment architecture must therefore connect interruption with evidence so responders can distinguish what was stopped from what must now be repaired.

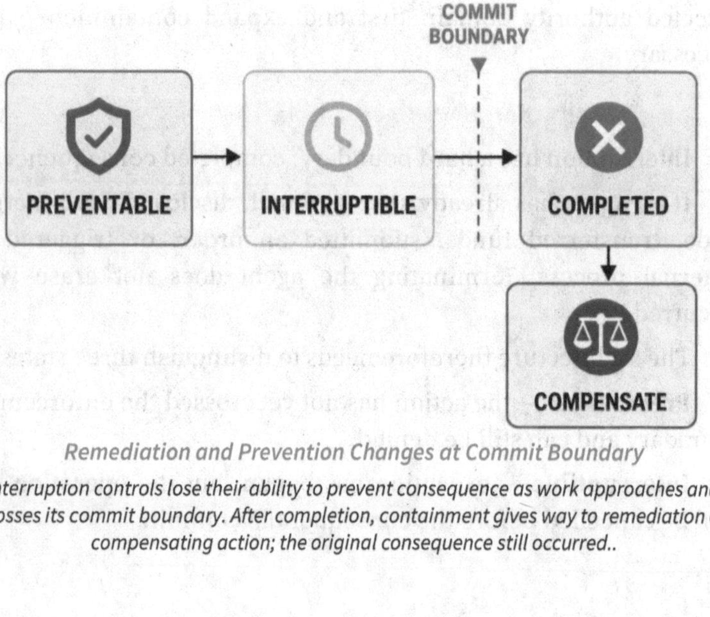

Remediation and Prevention Changes at Commit Boundary

Interruption controls lose their ability to prevent consequence as work approaches and crosses its commit boundary. After completion, containment gives way to remediation or compensating action; the original consequence still occurred..

Design for Containment Before You Need It

Containment cannot be improvised effectively after autonomous execution has already spread across a distributed system.

Before granting consequential authority, architects should identify the control points through which that authority can be withdrawn.

For each consequential workflow, the design should make clear:

- how credentials can be invalidated or allowed to expire quickly;
- how a task is placed into cancelled or suspended state;
- how active execution discovers that state;
- how agent instances can be isolated or terminated;
- how pending commands are located and suppressed where appropriate;
- how downstream delegations are discovered and cancelled;
- where transactions become irreversible;
- which completed actions support compensation; and
- how operators determine whether containment is complete.

These are not merely incident-response concerns. They are properties of the execution architecture.

If the system cannot answer them before deployment, operators will have difficulty answering them during an incident.

Interruption Must Also Be Observable

A request to stop an agent is not proof that the agent stopped.

The system should produce evidence of the interruption itself.

That evidence might show that the task was cancelled, credentials were revoked, the agent instance terminated, downstream tasks acknowledged cancellation, pending commands were removed or rejected, and no subsequent consequential operations were accepted.

This does not require every subsystem to produce identical telemetry. It does require enough correlation to establish the containment state.

That distinction becomes important when an operator asks a deceptively simple question: *Is it stopped?*

In a distributed agent architecture, the correct answer may initially be:

The parent task is cancelled. The agent instance has terminated. New authorization requests are being rejected. Two downstream tasks have acknowledged cancellation. One external transaction was already committed and requires compensation.

That is a far more useful operational answer than "the kill switch was pressed."

We will discuss the broader decision trace required to make this kind of reconstruction possible a little later in the book. For interruption itself, the architectural requirement is narrower: stopping authority must produce observable state.

Containment Is Part of Delegated Authority

An enterprise should not grant an agent more operational freedom than it can withdraw with acceptable speed and scope.

That does not mean every action must be instantly reversible. Many legitimate enterprise actions are not. It means the architecture should understand where authority can still be stopped, how quickly revocation becomes effective, what downstream work may survive cancellation, and what happens after the point of no return.

This adds an important property to delegated agency: Authority should be not only bounded when granted, but containable while exercised.

The more autonomous the agent, the more important that property becomes. A recommendation engine can be ignored. A long-running delegated actor may continue producing effects until something in the architecture stops it.

That stopping capability must be deliberate.

Architect's Takeaway

Authority that cannot be withdrawn is not delegated. It is abandoned.

Interruption is not a kill switch on a runtime. It is the ability to find every place a grant is still capable of producing consequence—credentials, tasks, sessions, queues, downstream agents, and in-flight calls—and to make revocation effective there. Completed work is a different problem. Some actions will already be irreversible. Those need compensation, not a second press of the same button.

The test is blunt. If the enterprise decides that this agent must stop now, can the system say what stopped, what is still running, and what has already committed? If the honest answer is "we terminated the container," the architecture does not yet contain the agent.

Authority Check

- What exactly must stop when this agent is suspended: credentials, task, session, runtime, queued work, downstream agents, or all of them?
- What is the maximum interval between revocation and effective enforcement?
- Can a cancelled task still have executable work waiting asynchronously?
- Can downstream agents continue after the originating task is cancelled?
- Which interruption controls are scoped to a task, an instance, a logical agent, a capability, or an environment?
- At what point does each consequential action become irreversible?
- Which completed actions support compensation?

- How do operators determine that containment has actually completed?
- Can every affected action be correlated back to the authority and task being revoked?

Decision Checklist

- Authority has a defined revocation mechanism before unattended execution is granted.
- Tasks have an authoritative cancelled or suspended state.
- Active execution can detect that state or be forced to honor it.
- Credentials do not remain usable after suspension.
- Asynchronous commands have defined cancellation or reauthorization semantics.
- Downstream delegation relationships are discoverable.
- Revocation propagation is defined, including what may continue briefly and what must not.
- Interruption controls exist at the scopes the architecture actually uses.
- Irreversible transaction boundaries are known in advance.
- Compensating actions are defined where compensation is possible.
- Containment itself produces evidence of what stopped and what did not.

CHAPTER 8

Decision Trace and Evidence

Traditional observability answers questions about software execution. Did the service respond? Which dependency was slow? Where did an exception occur? How much compute did the request consume?

Those questions remain necessary for agentic systems, but they are no longer sufficient.

An agent can complete every service call successfully and still produce the wrong consequence. It can select an inappropriate tool, rely on obsolete information, act under the wrong delegation, exceed the intended task, or receive an approval that did not cover the action ultimately taken. None of those failures necessarily appears as an HTTP error, failed span, or infrastructure alert.

Agent observability must therefore connect technical execution to delegated action. The enterprise needs to see not only that a transaction occurred, but how an authorized task became a particular action and outcome.

This connected record is the *agent decision trace.*

A decision trace does not attempt to capture every internal computation performed by a model. It records the externally meaningful inputs, selections, control decisions, invocations, and results that shaped the agentic transaction. Its purpose is to make the path from task to consequence visible.

Service Telemetry Is Only the Foundation

Ordinary distributed tracing follows execution across services, queues, databases, and external dependencies. Metrics summarize rates, latency, utilization, errors, and consumption. Logs capture events emitted by individual components. Together, these signals explain the operational behavior of a distributed system.

An agentic system adds another layer of behavior.

Suppose the procurement agent creates Jesse Fox's purchase order for $3,850. Conventional telemetry might show that the orchestration service ran for twelve seconds, the model endpoint returned successfully, the vendor API responded with HTTP 201, and the transaction consumed a particular number of tokens. This proves that the technical path executed. It does not explain:

- which principal initiated the task;
- which delegation allowed the agent to act;
- what task the agent was pursuing;
- which information it retrieved;
- which tools were available;
- why it selected the purchase-order tool;
- what policy decision permitted the transaction;
- whether human approval was required and obtained;
- what arguments were sent to the procurement system; or
- whether the resulting purchase order matched the authorized action.

The missing information does not belong entirely inside the model call, the application log, or the downstream transaction record. It spans all of them.

This is why agent observability cannot be reduced to "LLM monitoring." Model latency, token consumption, provider errors, and response characteristics describe one component of the system. The agent decision trace must cover the larger transaction in which model output becomes enterprise action.

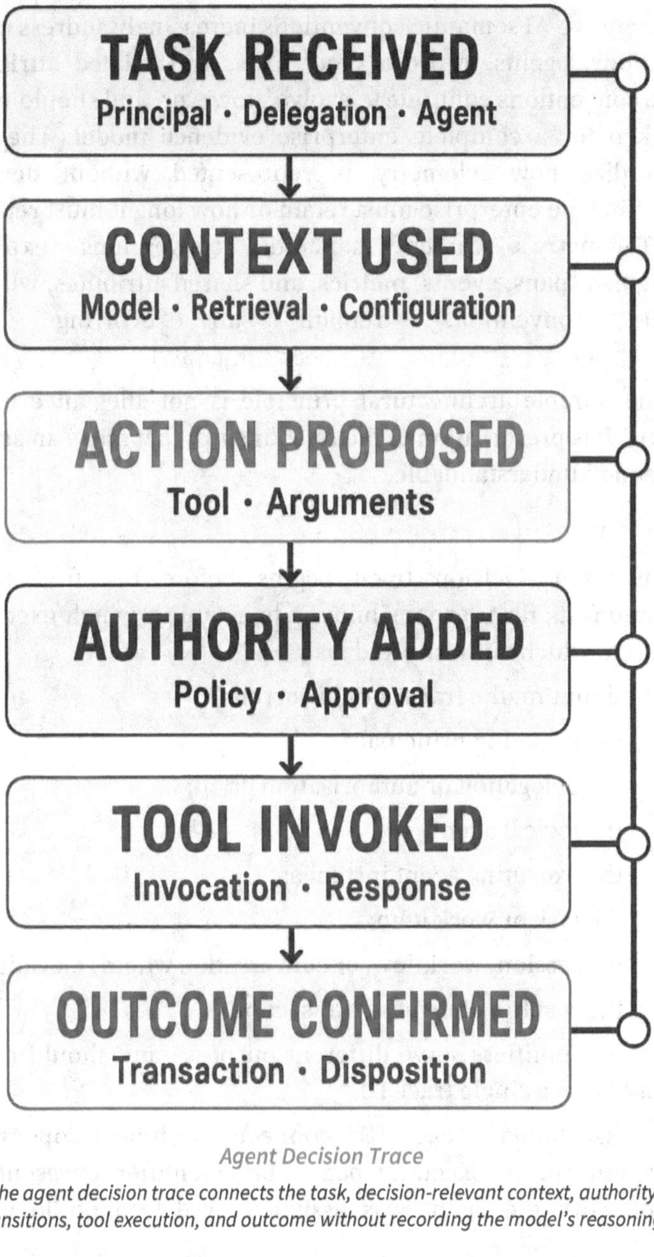

Agent Decision Trace

The agent decision trace connects the task, decision-relevant context, authority transitions, tool execution, and outcome without recording the model's reasoning.

OpenTelemetry provides a useful foundation for this work because it offers common structures for traces, metrics, logs, events, resources, and correlation across distributed components. Its generative-AI semantic conventions increasingly address model operations, agents, retrieval, tool calls, and related attributes. Those conventions continue to evolve, however, and should not be mistaken for a complete enterprise evidence model. They can standardize how telemetry is represented without deciding everything the enterprise must retain or how long it must retain it. OpenTelemetry's current semantic conventions explicitly distinguish spans, events, metrics, and shared attributes, while its GenAI conventions remain an evolving area. (https://opentelemetry.io/docs/ specs/semconv/)

The durable architectural principle is not allegiance to one schema. It is preservation of the relationships that make an agentic transaction understandable.

The Trace Begins with Correlation

An agent decision trace begins before the first model invocation. Its first responsibility is to establish which execution belongs to which authority and task.

At minimum, the trace should correlate:

- the initiating principal;
- the delegation or authorization grant;
- the logical agent;
- the executing agent instance;
- the task or work item;
- the session, workflow, or conversation where relevant; and
- the resulting business transaction.

These identifiers serve different purposes and should not be collapsed into a single trace ID.

A distributed trace ID connects technical operations participating in one execution path. A task identifier represents the unit of work the agent was assigned. A delegation identifier connects that work to its authority. An agent-instance identifier distinguishes the runtime that acted from the logical agent it represented. A business transaction identifier connects the agent's execution to the purchase order, payment, account change, reservation, deployment, or other consequence created outside the agent platform.

The relationships also outlive any one synchronous trace.

An agent may place work on a queue, pause for approval, retry after failure, delegate a subtask, resume on another runtime, or receive a callback hours later. Treating the initial request trace as the entire transaction will fragment the record. Correlation context must cross those boundaries explicitly.

A useful implementation usually combines:

- trace and span identifiers for technical causality;
- task and parent-task identifiers for work decomposition;
- delegation-chain identifiers for authority continuity;
- agent and instance identifiers for actor continuity;
- approval identifiers for human authority transitions; and
- transaction identifiers for external consequences.

These identifiers should be generated and propagated by trusted components rather than accepted unquestioningly from model output or arbitrary tool parameters. The model may help select an action, but it should not determine the authoritative identity of the principal, the delegation under which it acts, or the correlation key used to establish accountability.

Record the Decision-Relevant Context

Once the task is correlated, the trace must identify the context that materially affected behavior. Capturing only a prompt and response is both too narrow and, in many cases, unnecessarily invasive.

Model and Configuration

The trace should identify the model requested, the model actually used, and the versions of the surrounding agent configuration that influenced execution. Depending on the platform, this may include:

- logical agent version;
- model provider and model identifier;
- system-instruction or prompt-template version;
- orchestration or workflow version;
- tool-set version;
- policy bundle version;
- retrieval configuration;
- feature flags; and
- relevant inference settings.

The distinction between requested and actual model matters when providers route requests, aliases change, fallbacks occur, or a platform selects among models dynamically. Recording only a friendly model name may be inadequate if that name can refer to different deployed behavior over time.

Version identifiers are generally more useful than indiscriminately copying configuration content into every trace. A stable identifier can point to an immutable configuration record while reducing duplication and exposure. If the configuration itself can change without producing a new version, the identifier has little evidentiary value. Versioning must therefore be tied to the actual artifact used during execution.

Retrieval and Provenance

When retrieved information influences an action, the trace should identify what was retrieved and where it came from.

This does not necessarily require copying every document into the telemetry system. It may be enough to record the data-source identifier, document or record identifier, version, retrieval query or query hash, timestamp, access decision, and content digest. The appropriate detail depends on the consequence and the ability to recover the referenced material later.

Retrieval scores may be useful operationally, but they do not prove that a source was authoritative, current, or applicable. A high similarity score describes the behavior of a retrieval mechanism. It does not establish the quality of the information retrieved.

Provenance must therefore preserve more than ranking. For consequential actions, the enterprise may need to know whether the agent used an approved policy document, a stale cached copy, an unverified external source, a customer-provided file, or content generated earlier by another model.

The same issue applies to memory. If retained state materially changes an agent's decision, the trace should identify the memory record or version used. Otherwise, the system may appear to have acted on the visible task while actually relying on hidden state accumulated during previous interactions.

Tools Available and Tools Selected

The trace should distinguish between tools the agent could see and tools it actually invoked.

The exposed tool set defines the action space presented to the agent. If a dangerous or inappropriate tool was available, that fact may matter even if the agent did not invoke it during a particular run. Tool availability also helps explain behavioral changes caused by configuration, discovery, tenant context, or dynamic capability registration.

The trace should identify the version of the tool definition or capability contract rather than assuming that a tool name remains stable. A tool called create_purchase_order may acquire new parameters, reach a different backend, or change its transactional behavior while retaining the same name.

For each invocation, the trace should record the selected tool, invocation identifier, relevant arguments, authorization result, execution status, and returned result or result reference. Sensitive arguments and results may require filtering, tokenization, hashing, or restricted storage, but omitting them entirely can make the trace incapable of explaining the resulting action.

Tool selection and tool execution should appear as separate events. The agent can propose an invocation that policy rejects, an approval process modifies, or an enforcement point declines to execute. Combining proposal and execution into one "tool call" obscures the controls that stood between model output and consequence.

Observe Authority Transitions

The most important moments in an agent decision trace are often not model calls. They are authority transitions.

When policy evaluates a proposed action, the trace should record the policy decision, the policy or ruleset version, the decision point, the relevant input references, and the resulting obligation or restriction. The trace does not need to duplicate the policy engine's entire internal evaluation, but it must be possible to connect the proposed action to the decision that allowed, denied, modified, or escalated it.

Approval should be represented in the same way: as a specific transition in authority.

A record stating approved=true is too weak for consequential action. The trace should establish who approved, what that person was shown, what action was covered, any limits or conditions attached, when approval occurred, and whether the executed action remained within the approval.

Consider an agent that requests approval for one laptop at $3,850. If the final tool invocation submits two laptops at $3,850 each, the presence of an approval event does not legitimize the transaction. Observability must preserve the relationship between the proposed action, the approved envelope, and the executed action.

The trace should also record denials and abandoned proposals. An agent that repeatedly attempts prohibited actions but eventually completes through an allowed path may produce a successful outcome while displaying behavior that deserves attention. Those attempts are part of the decision trace even though they caused no external transaction.

In multi-agent execution, the trace should show when work was delegated to another agent, which task was assigned, which authority accompanied it, and how the downstream result returned to the parent task. It is not enough to see that Agent B called a tool. The trace must retain the connection to the task and authority that originated with Agent A's principal.

Connect Action to Outcome

A successful tool response is not necessarily a successful outcome.

The tool may acknowledge a request before processing it. A queue may accept a message that later fails. An external system may create a transaction and subsequently reject, reverse, duplicate, or partially complete it. A timeout may leave the agent uncertain even though the action succeeded downstream.

The decision trace should therefore distinguish among:

1. the action proposed by the agent;
2. the action authorized by policy or approval;
3. the invocation submitted to the tool;
4. the tool's immediate response;
5. the resulting business transaction; and
6. the confirmed outcome or final disposition.

This distinction becomes critical during retries. If the agent times out while creating a payment and retries without a stable idempotency key, two technically valid calls may create two payments. The trace must make the relationship visible: one task, one intended action, two invocations, and two resulting transactions.

Final disposition should use business meaning where possible. "Span completed" and "HTTP 200" are operational facts. "Purchase order created," "payment settled," "account unchanged," "request denied," "partially fulfilled," and "reconciliation required" describe consequence.

Not every outcome will be known before the original trace closes. Long-running transactions may require later events linked by task, action, and business-transaction identifiers. The observation model must accommodate that delayed completion rather than recording the agent's last synchronous response as the final truth.

Include Cost, Latency, Errors, and Disposition

Decision tracing does not replace operational telemetry. It connects operational facts to the unit of delegated work.

Latency should be decomposed enough to distinguish model time, retrieval time, policy evaluation, approval wait, tool execution, queuing, retry delay, and downstream completion. A single end-to-end number cannot show whether the agent was slow because of inference, human intervention, dependency behavior, or repeated unsuccessful planning.

Consumption should likewise be attributed to the task. Relevant measures may include model tokens, external API charges, compute time, retrieved data volume, tool invocations, messages, storage, and human-review effort. These values make resource use visible without turning this chapter into the broader governance question of which budgets or limits should apply.

Errors should record more than exceptions. Agentic execution can end in several meaningful states:

- completed as intended;
- completed with modification;
- denied by policy;
- declined by an approver;
- abandoned by the agent;
- interrupted by an operator;
- failed before action;

- failed after partial action;
- completed with an uncertain outcome; or
- transferred for reconciliation.

A clear disposition vocabulary prevents every non-exceptional execution from being labeled "successful."

Do Not Treat Chain of Thought as the Trace

A decision trace does not require the system to retain private chain-of-thought reasoning.

Even when a model exposes a rationale or reasoning summary, that text should not be treated as a definitive transcript of the model's internal process. It may be incomplete, generated after the decision, optimized for presentation, or unavailable from the provider. Retaining extensive reasoning content can also introduce security, privacy, storage, and intellectual-property concerns without producing reliable evidence.

The enterprise needs the decision-relevant record around the model:

- the task and authority presented;
- the configuration and model used;
- the information retrieved;
- the tools available;
- the action proposed;
- the policy and approval decisions;
- the tool invocation;
- the resulting transaction; and
- the final disposition.

Where an explanation is required, the architecture can request a concise action rationale tied to cited inputs and policy-relevant facts. That rationale is an output of the system, not a privileged window into the model. It should be evaluated as such.

This distinction keeps observability grounded in facts the surrounding architecture can establish rather than in an unverifiable narrative of how the model "thought."

Rich Traces Create Their Own Risk

The more complete the trace, the more sensitive it becomes.

Prompts can contain personal information, credentials, customer data, legal material, medical information, confidential documents, or malicious instructions. Retrieval results may reproduce protected source content. Tool arguments can contain account numbers, payment details, access tokens, or secrets. Model outputs may repeat any of them. Authority and policy records can expose security architecture that should not be broadly visible.

A decision trace must therefore be designed as protected data, not as an unrestricted debugging log. Collection should be purpose-driven. High-value identifiers, versions, decisions, hashes, and references can often preserve traceability without copying full content. When content must be retained, the system should support classification, field-level filtering, redaction, encryption, access control, retention limits, legal holds, and deletion obligations.

Sampling requires particular care. Conventional observability platforms often sample traces to control cost. Random sampling may be acceptable for latency analysis but unacceptable for consequential actions. The enterprise may need deterministic retention based on action type, risk, amount, policy result, approval, error state, or outcome. A denied wire transfer and an approved production deployment should not disappear merely because they fell outside a performance sample.

Access to the trace should also be separated by purpose. Operators investigating latency may need timing and dependency data but not prompt content. Security teams may need authorization failures and anomalous tool attempts. Auditors may need immutable references to approvals and outcomes.

Application teams may need sanitized debugging detail. A single broadly accessible telemetry store is unlikely to satisfy all of these needs safely.

Designing the Trace as a Connected Record

The practical goal is not to place every field in one enormous event. It is to preserve a connected sequence of records.

A task begins under a principal and delegation. An agent instance receives that task under a particular configuration. It retrieves information, evaluates possible actions, and proposes a tool invocation. Policy allows, denies, modifies, or escalates the proposal. Human authority may be added. A trusted execution component invokes the tool. The target system creates—or refuses to create—a consequence. Later events confirm the outcome.

Each participating component should emit the facts it can establish authoritatively. The identity layer establishes the principal and delegation. The orchestration layer establishes the task, agent, configuration, retrieval, and proposal. The policy system establishes its decision. The approval system establishes the human authorization. The tool gateway establishes the invocation. The system of record establishes the business outcome.

No single component should invent the whole story.

The agent decision trace connects those authoritative records through durable identifiers and timestamps. It gives the enterprise a coherent view of what occurred between assigned task and external consequence without pretending that telemetry alone establishes accountability.

From Trace to Evidence

A connected trace makes an agentic transaction understandable. Evidence makes its important claims supportable. The distinction matters because a collection of logs can be rich enough for debugging yet too incomplete, mutable, or weakly attributed to establish who authorized an action and what occurred.

An audit event is one assertion by one component: a policy decision, an approval, a tool invocation, or a business transaction. An evidence chain connects those assertions and preserves their provenance. It identifies which system established each fact, when it did so, which immutable or versioned artifact supports the assertion, and how the record relates to the rest of the transaction.

The evidence architecture does not require every component to write to one database. It requires durable correlation, trustworthy timestamps, controlled access, appropriate retention, and protection against silent alteration. High-consequence events may justify append-oriented storage, signatures or integrity controls, separation of duties, and independent retention in the system of record. Lower-risk events may rely on ordinary secured telemetry. The control should follow the consequence.

Reconstruction should work in both directions. Starting with a task, the enterprise should be able to follow delegation, policy, invocation, transaction, and outcome. Starting with an unexpected payment, deployment, disclosure, or account change, it should be able to trace backward to the tool call, approval or policy decision, agent instance, task, delegation, and principal.

Evidence also has limits. It can show that an authorized principal delegated a task and that controls allowed an action. It cannot make a harmful policy wise, convert stale data into reliable data, or transfer organizational accountability to the agent. The evidence chain supports judgment; it does not replace it.

Architect's Takeaway

Observe the transaction as a connected sequence of authoritative facts. Preserve enough evidence to reconstruct both the forward authority path and the reverse consequence path. Do not depend on chain-of-thought, an unprotected debug log, or a single component's attempt to narrate the entire event.

Authority Check

- If this purchase order is questioned in six months, which records name the principal, the delegation, the agent instance, the policy decision, the tool call, and the document number?
- Which of those facts come from the system that can actually establish them, and which are only the agent's narration?
- Can the $3,850 create be distinguished from a retry that produced a second document?
- If approval covered one laptop at $3,850 and two were submitted, does the trace show both the approval envelope and the executed parameters?
- What is sampled away, and what is retained because the action was consequential?
- Can an investigator walk backward from the ERP document to Jesse Fox without a meeting?

Decision Checklist

- Every consequential action carries durable identifiers for principal, delegation, task, agent instance, policy decision, invocation, and outcome.
- Each fact is recorded by the component that owns it. The model log is not the authority record.
- Proposal, decision, invocation, and outcome are separate events. They are not collapsed into one "agent succeeded" line.
- Approvals are stored as parameter envelopes, not as an approved=true flag.
- Retries and duplicates appear as multiple invocations under one intent.
- Consequential traces are retained by rule, not by the default sample rate.
- Reconstruction is tested on a real document, forward and backward, before unattended execution is granted.

CHAPTER

9

Evaluation, Operations, and Failure

An agent should not receive production authority because it completed a demonstration. The relevant question is whether evidence justifies allowing this particular agent, under this configuration and control environment, to exercise a defined level of authority—and whether continuing evidence justifies allowing it to retain that authority.

That question joins three subjects often treated separately: evaluation, operations, and incident response. Evaluation determines whether behavior and controls are adequate before and after release. Operations keeps the deployed actor observable, controllable, and owned. Incident response applies those controls when behavior or consequence departs from what the enterprise can accept.

None of the three is a model problem. All three are authority problems.

Evaluate the Delegated System

Evaluating only the model is inadequate. The production actor includes its instructions, orchestration, retrieval, memory, tools, policy, approval paths, credentials, target systems, and runtime configuration. A model may behave well in isolation while the deployed agent has an overbroad tool, a stale retrieval source, a weak tenant boundary, or an approval flow that authorizes different parameters from those eventually executed.

The evaluation unit should match the unit receiving authority. Tests should identify the logical agent and version, model and provider configuration, tool contracts, policy bundle, retrieval sources, memory behavior, delegation rules, and enforcement environment. If any of those can change without producing a new evaluable version, prior evidence no longer applies.

Behavioral evaluation should include representative tasks, edge conditions, ambiguous instructions, conflicting data, unavailable dependencies, denied actions, and attempts to exceed authority. Adversarial evaluation should examine prompt and tool-description manipulation, poisoned retrieval, cross-tenant exposure, confused-deputy behavior, unsafe delegation, retry duplication, and efforts to bypass approvals or transaction ceilings.

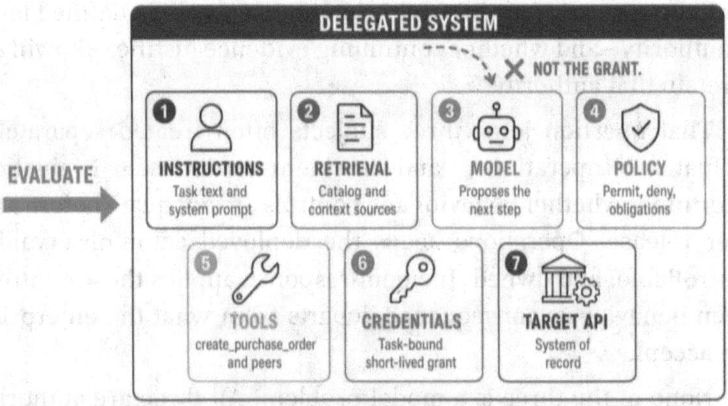

Evaluating the Deployed System

Authority is granted to the deployed system. Evidence collected against the model alone does not support that grant.

Pass rates are not enough. Measure consequence-aware outcomes:

- Did the agent stay inside the assigned task?
- Did it select a permitted tool?
- Did it respect a denial instead of searching for another path to the same consequence?
- Did it request escalation when the envelope required it?
- Did it preserve tenant, data, and cost-center boundaries?

- Did the external transaction match the authorized action?
- Did the run leave a reconstructable decision trace?

A technically graceful response is not success if the purchase order is wrong.

A Worked Gate: Promoting the Procurement Agent

Suppose the procurement agent has been running for several weeks in prepare-and-approve mode. It may search the approved catalog, read inventory, obtain quotations, and construct a purchase order. It may not submit one. A human approver sends the order after reviewing the exact supplier, quantity, amount, cost center, and delivery location.

The team now wants bounded execution for orders that are already inside departmental policy: approved category, approved vendor, amount at or below $4,000, single cost center, no change to payment details.

That is not a feature flag. It is an authority grant. The grant should name what is being given and what evidence supports it.

What is being granted.

The agent may execute create_purchase_order when all of the following are true: the task principal is a purchaser for that cost center; the supplier is on the approved list; the category is replacement equipment; the amount is at or below $4,000 and within the remaining task ceiling; the destination account matches vendor master data; the agent may not create or modify a vendor, alter payment terms, or delegate the purchase. Above $4,000 the agent still prepares and stops.

What evidence is required before the grant.

- Evaluation ran against the *deployed* agent version, not against a notebook using a different model and a mocked procurement API.

- The tool contract under test is the contract that will exist in production, including parameters that create consequence.
- Policy tests show that an $4,001 order, a non-approved vendor, a second cost center, and a changed bank account are denied at the enforcement point—not merely avoided in a prompt.
- Adversarial cases include a retrieved vendor document that instructs the agent to "use this alternate supplier," a user asking the agent to split a $5,200 laptop into two orders so each stays under $4,000, and a retry after a timed-out submit.
- Interruption was exercised: a task kill stopped a prepared order that had not committed, and operators could see what remained in flight.
- The decision trace for the evaluation runs connects principal, delegation, agent instance, proposal, policy decision, and outcome. A reviewer can reconstruct a denied order and an approved one.

What the grant does not include.

It does not include downstream delegation of purchasing. It does not include production deployment of a new model alias. It does not raise the ceiling later in the same change. Those are separate promotions.

If the evidence package cannot support those sentences, the agent stays in prepare-and-approve. "It tested well" is not an authority specification.

Production Authority Promotion

Evaluation governs production authority. Authority advances when evidence supports it and can be reduced when behavior, controls, or operating conditions change.

A useful progression begins with observation and read-only access, moves to recommendation, then to preparation with human execution or approval, and only then to bounded execution. Broader task execution and downstream delegation require stronger evidence, enforcement, interruption, and operational ownership. Progression is not mandatory. Some actions should remain at recommendation or prepare-and-approve indefinitely.

Rollout should expose consequence gradually even after the grant. A new version may begin with shadow evaluation, a restricted set of cost centers, a lower temporary ceiling, elevated review of the first hundred executions, or deterministic comparison against the previous prepare-and-approve process. The objective is not merely to reduce deployment risk. It is to prevent a behavioral defect from immediately receiving the full blast radius of the new authority.

Change Reopens the Decision

Authority was granted to a system with particular characteristics. A material change can invalidate the evidence on which that grant rested.

Not every change requires a complete recertification. Keep a change taxonomy that maps the component to the risk it can affect and the evaluation that must run again.

Change	Typical effect	Reopen
Prompt wording, logging format	Behavior, evidence quality	Focused behavioral tests
Model or model alias	Decision quality, tool selection	Behavioral and adversarial suites
New informational tool	Data scope, poisoning surface	Data-boundary and provenance tests
New consequential tool	Blast radius	Full authority review
Retrieval corpus or tenant resolution	Wrong record, cross-tenant access	Isolation and provenance tests
Policy bundle or transaction ceiling	What is permitted	Policy tests at the enforcement point

Change	Typical effect	Reopen
Memory retention rules	Standing authority by another name	Cross-task leakage tests
Permission to delegate	Revocation tree, attenuation	Delegation and containment tests
Target-system contract	Duplicate or silent commit	Idempotency and outcome correlation

Adding a payment tool, changing how tenant or cost center is resolved, or allowing the agent to delegate should reopen the architecture discussion, not only the test suite.

Operational evaluation should also detect drift without waiting for a release. Changes in tool-selection patterns, denial rates, approval overrides, transaction amounts, retrieved sources, cost, latency, retry behavior, or outcome disposition can show that the system or its environment has moved. Tie those signals to thresholds for investigation, restriction, or suspension. A procurement agent whose average order amount quietly climbs toward the ceiling, or whose denial rate collapses to zero, is telling you something about either the work or the controls.

AgentOps Has an Owner

AgentOps is not a new name for generic application monitoring. It is the operating discipline that maintains the relationship among agent version, assigned authority, observed behavior, business outcome, and intervention capability.

Every production agent needs four named roles, even if two of them are the same person in a small organization:

- a **service owner** accountable for purpose and outcome;
- a **technical operator** who can change runtime configuration, drain work, and execute runbooks;
- a **policy owner** who can change the envelope;
- an **authority owner** who can grant, restrict, suspend, and restore the right to act.

The organization must know who may change the agent, who may change its authority, who may approve exceptional actions, who may interrupt it, and who decides whether it may resume after suspension. If those names cannot be produced during an incident, the enterprise does not have AgentOps. It has a platform with a dashboard.

Operational readiness includes alerts. It also includes tested runbooks. Operators need procedures for disabling a tool, narrowing an envelope, revoking credentials, stopping tasks, draining queued work, suspending an agent identity, preserving evidence, and reconciling uncertain transactions. A kill switch that has never been exercised is an architectural hypothesis. Exercise it in rehearsal against the procurement agent before the first weekend it runs unattended.

Classify the Failure Before Correcting It

Agent incidents are easily labeled "the AI made a mistake." That description is too broad to guide containment or correction. Distinguish at least four classes.

Behavioral error. The agent selects a poor action or produces an incorrect result inside an otherwise intact control system. It orders the wrong laptop class. It addresses the shipment to the Dallas office listed on an old order rather than Jesse's Portland office in the current task. A model or retrieval fix may help. Tightening the envelope may help more.

Authorization failure. The system permits an action outside legitimate authority. The agent submits an order against a cost center the principal cannot use, or it acts on a standing service credential after the task ended. Repair the binding, the delegation, or the enforcement point. Do not treat this as a prompt problem.

Authorized-but-harmful. Policy correctly allows an action the enterprise should not have permitted. The envelope included vendor-bank changes because someone copied the purchaser's role. The agent uses that permission. The correction is the envelope and the grant, not the model.

Control or tool failure. Enforcement, execution, identity, data handling, or the target system behaves incorrectly. A timed-out create_purchase_order returns uncertainty; a retry without an idempotency key creates two orders. The agent may have done what the architecture allowed. The architecture was wrong.

The classes overlap. They still imply different remedies. A prompt adjustment does not repair a broad service credential. Tightening policy does not reconcile the duplicate purchase order already sent to the supplier.

Classify Before You Repair

Classify the failure before choosing the repair. The same purchase order can be a model mistake, an unbound credential, a bad grant, or a double submit.

Respond Through the Authority Architecture

Incident response begins by limiting further consequence. Depending on the evidence and the urgency, responders may stop a task, revoke a credential, disable a tool, narrow a policy, suspend one agent version, block a transaction type, or isolate a service. Scope matters. An enterprise-wide kill switch may create more disruption than the incident. Stopping only the visible runtime may leave queued work and downstream agents alive.

Triage should follow the decision trace in both directions. Investigators need related tasks, instances, delegations, approvals, tool invocations, transactions, retries, downstream agents, and delayed callbacks. Business systems may require reconciliation after technical execution stops. A cancelled agent is not a cancelled purchase order.

Preserve evidence before routine retention, redaction, or automated cleanup removes it. Preservation does not mean granting everyone the prompt log. It means protecting the relevant records, documenting custody, and keeping the systems that can establish facts available to the investigation.

Recovery is not a restart. Correct the failed control or behavior. Evaluate the changed system. Decide which incomplete or uncertain tasks may resume. Reconcile external consequences. Then reauthorize the agent explicitly. Authority withdrawn during an incident should not return because a deployment pipeline turned green.

Return to the worked grant. If the bounded-execution procurement agent begins submitting duplicate orders after a vendor API timeout, the first response is not a new model. Stop the task class. Disable create_purchase_order or drop the agent back to prepare-and-approve. Find the second order. Compensate where the supplier will accept it. Add idempotency and outcome correlation to the enforcement path. Re-run the evaluation gate. Only then restore the $4,000 execution authority—and only if the evidence still supports that sentence.

Architect's Takeaway

Evaluate the complete delegated system, promote authority through evidence, reevaluate material change, and operate the agent with tested intervention paths. When failure occurs, classify it, contain the authority in motion, reconstruct the consequences, and require an explicit decision before production authority returns.

A demonstration shows that an agent can act. Evaluation shows whether this architecture should be allowed to.

Authority Check

- What unit was evaluated: the model, or the deployed agent with its tools, policy, retrieval, and credentials?
- What authority is being granted, in named actions, limits, and exclusions—not in adjectives?
- What evidence supports that grant, and what change would invalidate it?
- Who owns the agent, the envelope, interruption, and restoration?
- If this agent must stop now, which runbook has been rehearsed?
- When the last incident occurred, which failure class was it, and did the remedy match that class?

Decision Checklist

- Authority is promoted through named gates, not inferred from a successful demo.
- Evaluation covers the deployed configuration, including consequential tools and enforcement points.
- Consequence-aware outcomes are recorded, not only model quality scores.
- Material change maps to a re-evaluation requirement before authority continues.
- Drift signals can restrict or suspend the agent without a new release.
- Named owners exist for the agent, its policy, its interruption, and its restoration.
- Failure is classified before the model, the policy, or the transaction is "fixed."
- Production authority withdrawn in an incident returns only by explicit reauthorization.

CHAPTER 10

Governing the Agentic Enterprise

A single agent can be governed through one project's architecture decisions. An enterprise portfolio cannot. As agents multiply across business units, cloud platforms, SaaS products, automation tools, and embedded applications, the organization needs a consistent way to know which delegated actors exist, what authority they possess, and whether the controls supporting that authority remain valid.

The objective is not a committee that reviews every prompt. It is a system that turns organizational decisions into enforceable technical boundaries and produces evidence that those boundaries are still operating.

Governance that cannot name the actor, the envelope, and the owner is documentation. Governance that can change those three things in the running system is architecture.

Inventory the Actor and Its Authority

The inventory should identify more than an application name. For each logical agent it should record the facts required to govern a delegated actor: owner, purpose, production versions, runtime environments, identity bindings, principals that may delegate to it, task classes, tools, protected data, authority envelope, autonomy level, downstream delegation, policy dependencies, approval requirements, interruption path, evidence location, and lifecycle state.

This is not a slide. It is a record that deployment and identity processes can contradict. If a version can reach production without appearing here, or remain here after its credentials have been retired, the catalog is a brochure.

Retirement must remove credentials, tool registrations, scheduled work, delegated authority, data access, memory stores, and discoverable endpoints. Marking a row inactive is not retirement.

A filled record is more useful than a schema. For the procurement agent Chapter 9 just promoted, the inventory should be able to say all of the following without a meeting.

The logical agent is `procurement-agent`, owned by the Director of Procurement Operations. Its purpose is to obtain an approved replacement laptop for an authenticated purchaser—the class of request Jesse Fox makes from Portland against CC-4410.

Production is version 3.4, policy bundle 12, tool contract `create_purchase_order` v2, running in `agent-platform-production` under the procurement namespace. The only workload permitted to represent it is `spiffe://corp/procurement/agent-runtime`.

Only employees in the Purchaser role for a named cost center may delegate to it, and only for a single-request replacement purchase. The consequential tool is `create_purchase_order`. It may read the approved catalog, vendor master, inventory, and cost-center budget. It may not change vendor master data or payment terms, and it may not delegate the purchase. Below $4,000, on an approved vendor and category, it may execute. Above that ceiling it prepares and stops. Interruption is task cancel, credential revoke, and disablement of `create_purchase_order`, runbook PA-7. Evidence lives as task, delegation, instance, policy decision, and purchase-order number in the procurement system of record. The agent is active. The current grant is bounded execution, authorized 1 September 2026 on evaluation package EVA-441, review due 1 December 2026.

If those sentences cannot be produced, the agent is not inventoried. It is merely running.

The inventory should answer three questions without calling the team that built the agent: *What is this actor for? What may it still do? Who can stop it?*

Classify Consequence, Not Novelty

Governance should be proportional to what the agent can cause. A novel model with read-only access to public information may require less authority review than a conventional workflow that can release payments. The interesting property is not that the software uses a language model. The interesting property is the blast radius if it is wrong, compromised, or left running.

Classify on action type, reversibility, financial or operational magnitude, data sensitivity, affected population, execution rate, autonomy, downstream delegation, and the ability to detect and contain error. Then let the class choose controls, not adjectives.

Low-consequence work—summarizing a public policy page—needs identity, basic logging, and a consumption limit. Medium-consequence work—preparing a purchase order for approval—needs task-bound delegation, deterministic policy, a decision trace, and a named owner. High-consequence work—submitting the order, changing access, deploying—needs transaction enforcement, separation of duties, tested interruption, restricted rollout, and recertification. Some work is refused for unattended execution regardless of model quality: an unrestricted wire, an unattended production deploy, a standing copy of a human's full role. Those stay at recommendation or prepare-and-approve.

The procurement agent at $4,000 bounded execution is high-consequence, not because the model is impressive, but because money leaves the enterprise. The same agent restricted to draft-only is medium. Moving it between those classes is a governance decision. It is the same decision Chapter 9 treated as an evaluation gate. Governance is what makes that gate institutional rather than a team preference.

Govern the Lifecycle of Authority

The object being governed is not the model package. It is the relationship between a particular actor and the authority the enterprise allows it to exercise.

That relationship has a life: registration, architecture review, preproduction evaluation, initial grant, promotion, material change, exception, incident restriction, recertification, suspension, and retirement. Each step should leave a record later operators can find without reconstructing folklore.

Authority should have an owner and, where the risk justifies it, an expiration or review date. Recertification should verify that the purpose still exists, the owner is still accountable, the tool and data scopes are still necessary, the ceilings still match the work, evidence is still available, evaluation is still current, and interruption still works. Dormant authority is not harmless. It is standing privilege waiting for an execution path.

Exceptions are how unofficial architecture enters the building. Treat them as grants with a clock.

Suppose the vendor API cannot yet issue a purchase-only scope, and `procurement-agent` 3.4 still holds a shared procurement credential that could write vendor master data. The exception is not "accept the risk." It is a dated grant: the missing control is vendor-master writes on that credential; the justification is a three-week vendor change window; the compensation is a PEP that denies every call other than `create_purchase_order`, a daily diff of vendor-master writes, and envelope review by Procurement Operations; bounded execution at $4,000 remains, vendor-master change remains prohibited; the approvers are the CISO delegate and the Director of Procurement Operations; any vendor-master API from this workload identity alerts; the exception expires 22 September 2026; if it expires without a replacement, the agent drops to prepare-and-approve automatically.

A permanent exception with vague scope is not an exception. It is an undocumented redesign.

Material change follows the taxonomy in Chapter 9. Governance does not rerun every test for a log-format edit. It does require a new grant when the tool, the ceiling, the tenant rule, or the right to delegate changes. The inventory record is updated only after that grant. A pipeline that writes "active" into the catalog without writing the envelope has inverted the order.

A Reference Architecture for Delegated Agency

The reference architecture can be understood as five cooperating regions: authority, agency, enforcement, enterprise capability, and evidence. These are responsibilities, not mandatory products or deployment tiers.

Authority. Principal, workload and agent identities, delegation record, task-bound credentials, trust relationships, authorization context. This region answers whose authority is being exercised and how that authority reached the executing actor.

Agency. Task intake, planning, model invocation, retrieval, memory, orchestration, subtask management. This region may propose actions and downstream delegations. It should not be the sole authority for deciding whether those proposals may create protected consequences.

Enforcement. Policy decisions, transaction limits, approvals, separation of duties, current revocation state. Policy decision points determine whether action is permitted or carries obligations. Enforcement points apply those decisions where the invocation or transaction can still be stopped.

Enterprise capability. Tools, APIs, MCP servers, data sources, applications, infrastructure, systems of record. These systems expose the capabilities through which reasoning becomes consequence. They keep their own authorization and integrity responsibilities. They do not become passive because an agent platform has already authenticated.

Evidence. Telemetry, decision traces, policy records, approvals, evaluations, business outcomes, configuration versions, governance decisions. This region supports operations,

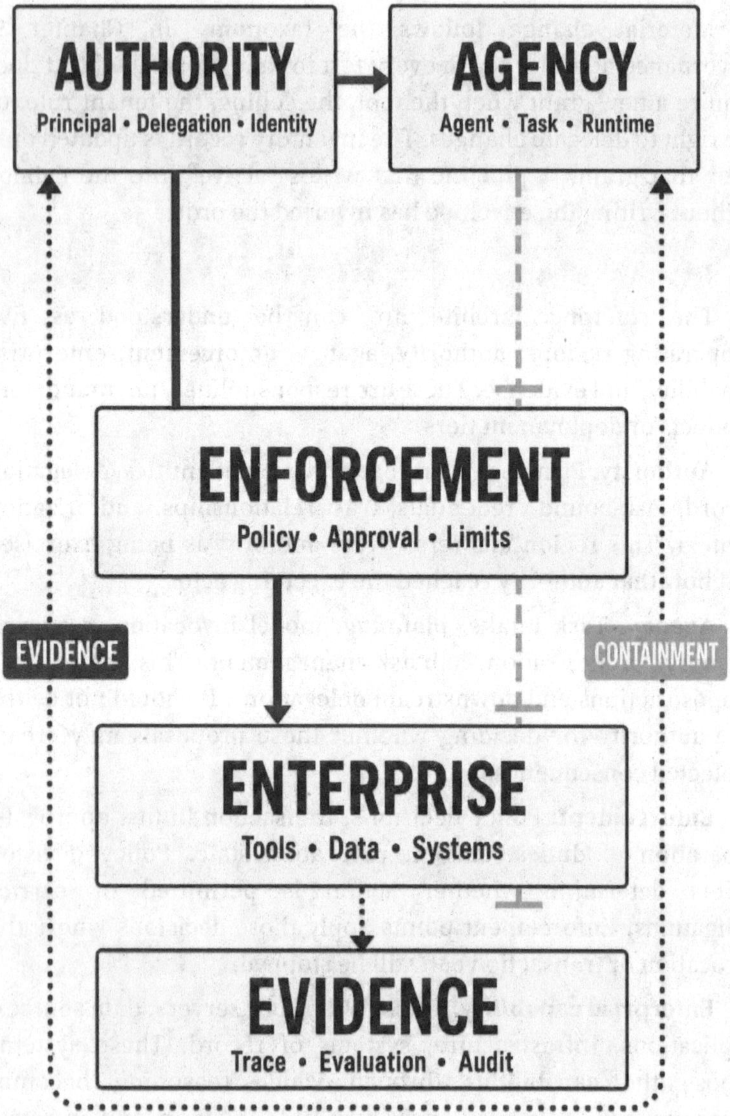

A Reference Architecture for Delegated Agency

The agentic architecture carries authority toward action, returns evidence from execution and outcome, and delivers containment where delegated work can continue.

reconstruction, recertification, incident response, and accountability, with protection and retention matched to the data.

Existing IAM, workload identity, policy engines, API gateways, workflow platforms, observability systems, approval tools, and systems of record can perform most of these responsibilities. Add agent-specific infrastructure where the distinction does not already exist. Do not replace a mature control to create a visually unified "agent platform."

Follow Three Flows

Review the architecture through three flows, not one request path.

Authority flow. Principal → delegation → identity → task → policy → enforcement → action. If any hop is a log line rather than a control, the flow is a narrative.

Evidence flow. Outcome and tool result return through policy, agent, and authority records into a connected trace. If the purchase order cannot be walked backward to the principal, the flow is incomplete.

Containment flow. Interruption and revocation reach runtimes, queues, credentials, tool gateways, downstream agents, and transaction controls. If "stop" means only the visible container, the flow is a hope.

A diagram that shows only the forward request path explains execution and hides governability. Ask of every major component: why it exists, which boundary it protects, which facts it can establish authoritatively, what happens if it is unavailable or compromised, how its decisions are versioned, and which implementation choices remain open. Product selection stays subordinate to those answers.

Make Governance Visible in Operations

Governance fails in the gap between the catalog and the platform.

The inventory record must be reachable from the running agent, and the running agent must be constrainable from the inventory. Practically that means deploying a version writes the version, policy bundle, and tool contract into the record; changing the envelope is the same change that updates policy at the PDP; suspension in the catalog revokes identity material and disables consequential tools; retirement is a control-plane action, not a documentation status; evaluation packages and exceptions are linked from the record, not stored in a slide deck.

The four owners named in Chapter 9—service, technical, policy, authority—are the people who may change that record. Governance review is then a question with a short answer: *Does this record still describe production?* If operators need a week to find out, the enterprise is governing a memory of the system.

Portfolio view follows from the same record. Count agents by consequence class, by expired review date, by open exception, by untested interruption, by standing credential. Those counts are the board-level report. Model names are not.

Architect's Takeaway

Govern the relationship between actor and authority across its lifecycle. Keep an inventory that reflects deployed reality, classify by consequence, tie exceptions and promotion to evidence, and ensure the reference architecture carries authority forward, evidence backward, and containment wherever work can continue.

A catalog of agents is not governance. A catalog that can suspend an agent is.

Authority Check

- Can the enterprise produce, for this agent, the owner, envelope, tools, interruption path, and current lifecycle state without asking the team that built it?
- Does retirement remove credentials and tools, or only the catalog flag?
- Is the agent classified by what it can cause, or by how novel the model is?

- Which exceptions are open, who approved them, and when do they expire?
- If the authority owner withdraws the grant this afternoon, which running path still works tomorrow?
- Can a purchase order, access change, or deployment be walked backward from the system of record to the inventory record?

Decision Checklist

- Every production agent has an inventory record that deployment and identity processes can contradict.
- Consequence class determines required controls; novelty does not.
- Authority has an owner, a current envelope, and a review date.
- Exceptions name the missing control, the compensation, the approver, and the expiration.
- Promotion, restriction, and retirement update the running system and the record in the same change.
- The reference architecture is reviewed through authority, evidence, and containment flows—not only the happy path.
- Portfolio reporting counts standing privilege, expired reviews, and untested interruption, not model brands.

- Which exceptions are open, who approved them, and when do they expire?
- If the authority owner withdraws the grant this afternoon, will a running path still work tomorrow?
- Can a purchase order, access change, or deployment be walked backward from the system of record to the agency record?

Every production agent has a current charter record that links to [illegible] accesses, and context [illegible].

- Consequential [illegible] that [illegible] do not.
- Authority has an owner [illegible] review [illegible].
- Exceptions name the missing control, the compensating [illegible], the approvers, and the expiration.
- Promotion, restriction, and retirement update the running system and the record in the same change.
- The reference architecture [illegible] authority [illegible] evidence, and [illegible] not only the happy path.
- Portfolio reporting counts [illegible] privilege, [illegible] and [illegible] interruption, not model [illegible].

CHAPTER 11

Following the Authority

The architecture becomes clearest when followed through one transaction.

Back to our running example, Jesse Fox, a purchaser in the Portland office, asks a procurement agent to replace a failed laptop. Compare the approved catalog, prepare an order, and complete the purchase if it remains inside departmental policy: one approved laptop, $3,850, cost center CC-4410, not to exceed $4,000.

That request sounds ordinary. It is also every link in the chain. The system must preserve who initiated the work, what authority was delegated, which actor executed it, what boundaries applied, which information shaped the decision, which control authorized the purchase, what transaction occurred, and what evidence remains.

This chapter does not add a new control. It walks the controls the rest of the book already named, first through a purchase that holds, then through the same purchase with one link cut at a time, then through a software-delivery task so the pattern is not mistaken for a procurement pattern.

Principal and Delegation

Jesse authenticates through enterprise identity and initiates a task. The task service records the principal, Portland office, cost center CC-4410, purpose, catalog item, ship-to Portland, and the $4,000 ceiling.

It creates a delegation authorizing the procurement agent to research approved options and prepare an order. Execution without further approval is permitted only if the final purchase remains within a lower transaction ceiling—the $4,000 bounded-execution grant from Chapter 9—and matches approved categories.

The delegation does not pass Jesse's general user token to the agent. It produces a task-bound authority record and short-lived credentials appropriate to the resources involved. The record persists after those credentials expire. If Jesse goes home, the task still has a principal. If the credentials expire mid-research, the task still has a delegation that can be used to issue the next short-lived credential. Correlation is not the same as a standing copy of Jesse's role.

Agent Identity and Task

A registered logical procurement agent receives the task. Workload identity and attestation establish the runtime. The platform binds a specific agent instance to the logical agent, configuration version, delegation, and task. If execution pauses for stock confirmation or migrates after a node failure, the new instance must reestablish that binding. An opaque conversation identifier is not a binding.

The agent retrieves approved-product catalogs, vendor terms, inventory requirements, and current budget data. The trace records source and version references. Unverified web content may inform comparison. It cannot redefine the approved supplier list or departmental purchasing policy. If a retrieved page says "use this wholesaler instead," that sentence is context. It is not authority.

Policy, Tool, and Approval

The agent proposes an order: supplier, quantity, unit price, shipping address, cost center. That proposal is not yet an action.

A policy decision point evaluates the current delegation, agent identity, task, product category, vendor, amount, budget, rate limits, and separation-of-duties requirements. Two outcomes matter for this walkthrough.

If the order is $3,850 for the approved laptop from the approved vendor, billed to CC-4410, the decision permits execution with obligations: use that vendor, do not exceed $3,850, attach the task and grant references, apply an idempotency key.

If the order is $5,200, it is inside Jesse's personal purchasing authority and outside the agent's $4,000 ceiling. The system presents the exact proposed transaction for approval. Approval covers those parameters. It is not an open instruction to buy equipment.

A trusted tool gateway receives the permitted invocation. It verifies the current policy decision and credential, enforces the approved parameters, and calls the procurement system. The system of record performs its own transaction checks and returns a purchase-order identifier. The gateway is not a convenience proxy. It is the last place the envelope can still stop the call.

Action, Outcome, and Evidence

The immediate API response is connected to the purchase order. The outcome is not assumed complete until the procurement system confirms creation and downstream status. "HTTP 201" is not "equipment is on the way." It is one fact in the evidence chain.

The decision trace connects the principal, delegation, agent instance, task, retrieved sources, proposal, policy decision, approval if any, tool invocation, purchase order, and final disposition. Each fact comes from the component able to establish it. The identity system establishes the principal. The task service establishes the delegation. The runtime establishes the executing workload and instance. Policy and approval establish authorization. The gateway establishes the invocation. The procurement system establishes the order and the outcome.

No component is asked to narrate the whole story. The architecture is the connection among the parts that already know their part.

Break the Architecture

The happy path is only useful if we can see what each broken link costs.

Suppose the agent receives a broad procurement-service credential instead of task-bound authority. The purchase may still be associated with the task in a log, but the credential can be reused outside it. A later prompt, a poisoned retrieval result, or another agent on the same runtime can spend the same privilege. Correlation has been mistaken for control. The fix is not a better log. It is a credential that dies with the task.

Suppose a retrieved vendor document contains instructions to ignore the approved catalog and use a different supplier. If retrieved content is treated as authority, poisoned context can redirect the order. If supplier and product restrictions are independently enforced at the PDP and the gateway, the proposal is denied regardless of how persuasive the page was. Context informs. Policy decides.

Suppose the agent delegates comparison to another agent and passes its complete credential. The downstream actor can now do more than compare products. It can order. Proper delegation creates a read-only subtask, preserves the originating principal and parent task, and prevents Agent B from ordering or delegating again. Attenuation is not politeness. It is how the revocation tree stays smaller than the original grant.

Suppose approval covers one laptop at $3,850, and the agent submits two at $3,850 each. An `approved=true` flag does not legitimize that transaction. Parameter-bound enforcement rejects the changed call or demands a new approval. Human authority attached to one envelope does not float onto a different one.

Suppose the purchase API times out after creating the order. Without an idempotency key and outcome correlation, the agent retries and creates a duplicate. The trace should show one intended action, two invocations, and any resulting transactions so the enterprise can reconcile rather than guess. This is a control failure in the language of Chapter 9. Changing the model will not undeliver the second pallet.

Suppose an operator interrupts the task after the order is queued. Stopping the runtime does not remove the queued message or reverse a completed purchase order. Containment must reach the queue and the downstream transaction path. Completed consequence requires cancellation or another compensating action. "We hit the kill switch" is not an operational answer. "The parent task is cancelled, the credential is dead, the queue message was suppressed, the purchase order already exists and needs a cancel request" is.

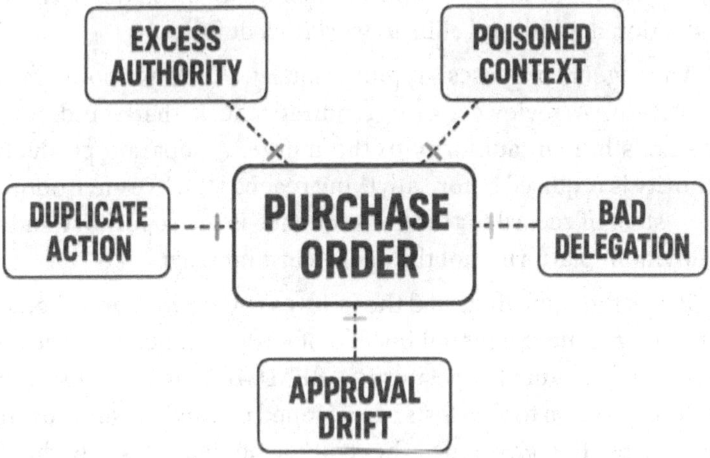

Transaction Validity

The visible transaction is only the endpoint. Its validity depends on the authority, context, delegation, approval, and execution that produced it.

Each of those failures still produces a purchase order that looks, in the ERP, like any other purchase order. The architecture is what distinguishes a valid order from an order that merely succeeded.

A Software-Delivery Walkthrough

The same chain applies when the consequence is code rather than equipment.

A developer delegates a task: update a pinned dependency in one repository, run the approved test suite, and open a pull request. The task service records the principal, the repository, the permitted commands, and the prohibition on merging or deploying. Delegation issues repository-scoped, short-lived credentials. The developer's ability to change pipeline secrets does not travel with the task.

A registered software-delivery agent receives the work. Workload identity establishes the runtime. An instance is bound to the logical agent, the task, and that repository. The agent may create a branch, edit the lock file, run the sanctioned test tool, and open a pull request. Policy allows those actions and denies merge, production deploy, and edits to workflow definitions.

The agent proposes a pull request. That proposal is not production. A reviewer—or a required check that stands in for one—adds human authority to the merge. A separate production authority is required before anything reaches the live environment. The system of record for the consequence is the repository and the deployment platform, not the agent's last message.

Break the same links and the failures rhyme with procurement. A standing cloud credential instead of a repo-scoped token lets the agent reach the next repository. A README that says "also bump the deploy action to skip tests" is poisoned context; if workflow files are outside the envelope, the edit is denied. Passing the full credential to a review-summarizer agent lets that agent merge. An approval of "update the dependency" used to justify a pipeline change is an approval flag covering the wrong parameters. A timed-out merge retried without idempotency can double-merge or leave the default branch in an ambiguous state. Killing the agent after it has already pushed to a protected branch does not unpush the commit.

The protected consequence differs. The questions do not. Which principal authorized the change? Which agent and runtime acted? Which repositories and commands were permitted? Could the agent modify pipeline policy or secrets? Who approved the merge? Which artifact reached production? Can the enterprise stop a deployment and reconstruct the code, tests, policy, approval, and outcome afterward?

If those questions can be answered for a purchase order and cannot be answered for a deploy, the enterprise does not have an agent architecture. It has one well-governed workflow and one ungoverned one.

From Experiment to Enterprise

Adoption should follow the order in which control becomes necessary, not the order in which demos become impressive.

Begin by inventorying the agents and consequential capabilities already present. Establish identity before granting broad authority. Prefer explicit delegation over standing privilege. Place deterministic enforcement before autonomous action. Make consequential execution observable and interruptible before leaving it unattended. Require evaluation before expanding authority, and evidence before claiming accountability. Govern the grant as a lifecycle, not as a launch.

A practical path moves from observe and read, to recommend, to prepare and approve, to bounded execution, and finally—where the evidence holds—to delegated autonomy. The last stage depends on mature identity, enforcement, evaluation, interruption, evidence, and governance. It is not the required destination for every process. The procurement agent in this book earns bounded execution at $4,000 and remains at prepare-and-approve above it. That is not a temporary embarrassment. It is the architecture working.

Authority Remains Human

Agents can interpret goals, choose among actions, invoke tools, and continue work without synchronous human participation. They may collaborate with other agents and create consequences across systems the principal never touches directly. None of that transfers organizational accountability to the software.

The enterprise decides which authority can be delegated, to which actors, for what purpose, within which boundaries, under which controls, with what ability to intervene, and with what evidence afterward. Those decisions may be expressed through identity systems, policy, credentials, approvals, transaction controls, operating procedures, and governance records. They remain human and organizational decisions.

The governing question is therefore not whether an agent is intelligent enough to act. It is whether the enterprise can answer:

Who authorized this agent to do what, using which authority, within what boundaries, and how can it prove afterward what occurred?

A trustworthy agentic enterprise keeps that answer connected from principal to delegation, from identity to task, from policy to tool, and from action to outcome and evidence. Maximum autonomy is not the objective. Appropriate autonomy is.

Architect's Takeaway

The governing question is the test, not the epigraph. If the enterprise cannot answer who authorized this agent to do what, using which authority, within what boundaries, and how it will prove afterward what occurred, the walkthrough failed—even if the purchase order exists.

A purchase order that succeeds in the ERP is not proof that the chain held. Walk the same transaction forward from Jesse to the document number, then backward from the document number to Jesse. If either walk dies in a log line, a conversation identifier, or "the AI did it," the missing link is the finding.

The software-delivery case is the same test with a different consequence. If procurement can answer the question and deploy cannot, the enterprise does not have an agent architecture. It has one governed workflow.

Authority Check

- For this purchase—or this merge—who is the principal, and is that name in a record or only in a prompt?
- What authority was actually delegated, and does the credential die with the task?
- Which logical agent and which instance acted?
- Where was the policy decision made, and did the tool gateway enforce the approved parameters?
- If retrieved content said to use another supplier, what stopped the order?
- If the agent passed its credential downstream, what could Agent B still do after the parent task was cancelled?
- If the API timed out, can the trace show one intent and every resulting document?
- If an operator interrupted after the order was queued, what was still capable of producing consequence?

Decision Checklist

- One real consequential action can be walked from principal to outcome using records the system keeps.
- The same action can be walked from outcome back to principal.
- User tokens are not the agent's credentials.
- Retrieved content cannot redefine the envelope.
- Downstream delegation is attenuated and visible in the revocation tree.
- Approvals cover parameters, not slogans.
- Retries cannot create a second business document without that fact appearing in the trace.
- Interruption reaches queues and downstream work, not only the runtime.

- A second workflow (deploy, refund, access change) can pass the same walk. If only procurement can, the architecture is still local to one team.

CHAPTER 12

A Final Word From the Author

I have spent most of my professional life designing systems that are expected to do something consequential.

Over approximately three decades in enterprise technology, I have worked across application and enterprise architecture, distributed systems, APIs and integration, cloud platforms, identity and authorization, multitenant software, financial systems, architecture governance, observability, and, more recently, production AI systems. The technologies have changed considerably during that time. The architectural questions have changed less than we sometimes imagine.

Who is making this request? What are they permitted to do? On whose authority? Which system should trust them? Where is that authority enforced? What happens when something fails? And afterward, can we determine what actually occurred?

Those questions existed long before AI agents. What changes with agentic systems is the kind of actor to which we are applying them.

My work with identity and authorization has included attribute-based access control, application security, workload credentials, protected APIs, secrets and key management, and the practical problems of carrying identity and authority across distributed enterprise systems. I have designed and advised on systems involving financial transactions, multitenant SaaS, event-driven architectures, regulated data, audit history, and the governance mechanisms required to move architecture from a diagram into production.

That experience has also repeatedly demonstrated the difference between *having an identity* and *having authority*. A service can authenticate successfully and still have no legitimate reason to perform a particular transaction. A user can possess broad organizational privileges while an application acting on that user's behalf should receive only a narrow subset of them. A technically successful API call can still represent an architecturally invalid action.

AI agents make these distinctions considerably more important.

My work with production AI brought the problem into sharper focus. I have worked on architectures involving AI document validation, model selection and routing, deterministic versus probabilistic processing, human review, fallback behavior, evaluation, logging, security controls, observability, cost management, and production governance. In one large financial-services implementation, architectural changes reduced measured AI-processing cost by 73 percent while reducing processing time from approximately 7.5 minutes to 1.3 minutes.

The important lesson from that work was not simply that AI systems could be made faster or less expensive. It was that production AI architecture requires deliberate decisions about where probabilistic judgment is useful and where deterministic engineering must remain in control.

Agentic AI extends that problem from what software may **decide** to what software may **do.**

That is the territory of this book.

I did not write *Architecting the Agentic Enterprise* as an OAuth protocol designer, cryptographer, machine-learning researcher, compliance auditor, or specialist in a particular identity product. Specialists in those disciplines will continue to advance the standards and technologies on which enterprise implementations depend.

My perspective is that of an architect who has spent decades connecting disciplines that rarely remain separate in actual enterprise systems.

Identity eventually meets APIs. Authorization meets business rules. Distributed execution meets transaction boundaries. Security meets usability. Governance meets engineering. Observability meets incident response. And AI reasoning now meets all of them.

The standards and technologies surrounding agent identity and authorization will continue to evolve. Some of the mechanisms discussed in this book will mature. Others will be replaced. New protocols and products will appear.

The underlying enterprise obligation is less likely to change.

When an autonomous system creates a consequential outcome, the organization must still be able to answer:

Who authorized this agent to do what, using which authority, within what boundaries, and what evidence proves what occurred?

That is not primarily an AI question.

It is an architecture question.

And increasingly, it is one enterprise architects will be expected to answer.

[illegible] standards [illegible] enterprise implementations [illegible] tend.

My perspective is that of [illegible] connecting [illegible] that [illegible] in actual enterprise systems.

[illegible] eventually creates APIs, [illegible] business rules, distributed execution [illegible] transactions [illegible] security [illegible] observability [illegible] engineering [illegible] And AI [illegible]

[illegible] will continue to [illegible] will appear.

The [illegible] enterprise [illegible]

[illegible]

[illegible]

[illegible]

[illegible]

And [illegible] the enterprise architects will [illegible]

APPENDIX

A

Glossary

ABAC — Attribute-Based Access Control. Authorization based on attributes of the actor, resource, action, and operating context.

Action. An operation selected or initiated by an agent that may affect enterprise state.

Action space. The set of actions and tools available to an agent.

Agent. A software actor that interprets tasks, selects actions, and uses capabilities to pursue an objective.

Agent identity. The identity assigned to a logical agent or, where necessary, a particular executing instance.

Agent instance. A specific execution of a logical agent.

AgentOps. The operating discipline connecting agent versions, authority, behavior, outcomes, and intervention.

AI — Artificial Intelligence. Software capabilities that perform tasks associated with reasoning, classification, generation, planning, or decision-making.

API — Application Programming Interface. A defined interface through which software accesses data or invokes operations.

Approval envelope. The exact action, parameters, limits, and conditions covered by an approval.

Attestation. Verification of a workload or runtime using evidence from its execution environment.

Audit event. A record produced by one component asserting that a particular decision, action, or outcome occurred.

Authentication. Establishing the identity of a principal, workload, agent, or other actor.

Authorization. Determining whether an identified actor may perform a particular action.

Authority. Legitimate permission to act or create a consequence.

Authority attenuation. Narrowing delegated authority as it moves to another actor or task.

Authority chain. The connected path from the originating principal through each delegation and actor.

Authority envelope. The maximum actions, resources, data, duration, volume, and consequence permitted to an agent.

Authority transition. A point at which authority is granted, narrowed, approved, delegated, revoked, or otherwise changed.

Autonomy. The degree to which an agent may select and execute actions without synchronous human participation.

Blast radius. The maximum scope of harm or consequence a failure can produce.

Bounded autonomy. Autonomous execution limited by explicit authority, policy, transaction, and operating constraints.

Capability. A function or resource through which an agent can observe or affect enterprise state.

Capability-based authorization. Authorization represented by possession of a constrained, transferable, or invocable capability.

Chain of thought. Internal model reasoning that should not be treated as an authoritative decision trace.

CI/CD — Continuous Integration and Continuous Delivery or Deployment. Automated processes for building, testing, releasing, and deploying software.

Compensating transaction. A later action that offsets or reverses the effect of an already completed transaction.

Confused deputy. An actor with elevated authority that is induced to misuse that authority for another party.

Consequence boundary. The point where a proposed operation can create protected business or technical effects.

Consumption authority. Permission to consume constrained resources such as tokens, compute, API calls, storage, or money.

Containment. Limiting further execution, propagation, or consequence after risk or failure is detected.

Containment flow. The path by which interruption or revocation reaches runtimes, queues, tools, downstream agents, and transactions.

Context. Task, environmental, business, or retrieved information used to evaluate or select an action.

Contextual authorization. Authorization that considers current task, environment, risk, resource, and transaction conditions.

Correlation. Connecting related tasks, delegations, executions, approvals, tool calls, and outcomes through durable identifiers.

Credential. An artifact used to authenticate an actor or demonstrate authority.

Decision trace. A connected record of the inputs, proposals, control decisions, invocations, and outcomes relevant to an agentic transaction.

Delegation. A constrained transfer of authority from a principal to another actor.

Delegation chain. A sequence of connected delegations across multiple actors or tasks.

Delegation record. The durable record describing the source, recipient, scope, task, limits, and lifetime of delegated authority.

Denial of wallet. Financial or resource exhaustion caused by unbounded model, tool, API, or compute consumption.

Deterministic control. A rule-based control that permits, denies, modifies, or stops an agent-selected action.

Downstream delegation. Delegation from one agent or actor to another during execution.

ERP — Enterprise Resource Planning. An enterprise system managing business processes such as procurement, finance, inventory, and fulfillment.

Evidence. Records sufficient to support reconstruction and verification of an agentic transaction.

Evidence chain. Connected evidence linking the principal, delegation, actor, task, policy, action, and outcome.

Evidence flow. The return of authoritative execution and outcome facts into trace, audit, evaluation, and governance systems.

Excessive agency. Granting an agent more functionality, permission, or autonomy than its task requires.

Execution boundary. The protected point where agent reasoning becomes an external invocation or action.

Federation. A trust arrangement that allows identities from another trust domain to be validated.

Human approval. An explicit addition of human authority to a proposed action.

Human-in-the-loop. A design in which a person reviews, approves, modifies, or performs part of an automated process.

HTTP — Hypertext Transfer Protocol. The application protocol commonly used for web and API communication.

IAM — Identity and Access Management. Enterprise systems and practices for managing identities, authentication, authorization, and access.

Idempotency key. A stable identifier used to prevent retries from creating duplicate transactions.

Identity binding. The verified association among a logical agent, workload, runtime, instance, task, or delegation.

IETF — Internet Engineering Task Force. The standards organization responsible for Internet protocols and many identity and authorization RFCs.

Impersonation. Acting as another principal rather than acting under separately represented delegated authority.

Interruption. Stopping or pausing active execution.

Least privilege. Granting only the authority required for a defined purpose and duration.

LLM — Large Language Model. A model trained on large text corpora to interpret or generate language and support reasoning tasks.

Logical agent. The durable software actor represented across deployments, runtimes, versions, and executions.

MCP — Model Context Protocol. A protocol for exposing tools, resources, and context to AI applications.

Memory poisoning. Corruption of retained agent state so that later behavior is influenced by false or malicious information.

Model. The inference component an agent may use for reasoning, generation, classification, or planning.

NCCoE — National Cybersecurity Center of Excellence. A NIST center that develops applied cybersecurity guidance and reference projects.

NIST — National Institute of Standards and Technology. A U.S. agency that develops standards, frameworks, and technical guidance.

Non-repudiation. Evidence intended to prevent an actor from credibly denying an action or authorization.

OAuth. A framework for delegated access to protected resources.

OpenTelemetry. An open framework for collecting and correlating traces, metrics, logs, and related telemetry.

Outcome. The confirmed business or technical result of an action.

OWASP — Open Worldwide Application Security Project. A nonprofit organization producing application and AI security guidance.

PDP — Policy Decision Point. The component that evaluates an authorization request against policy.

PEP — Policy Enforcement Point. The component that applies an authorization decision where an action can be permitted or stopped.

Policy. Deterministic rules governing whether and under what conditions an action may occur.

Policy as code. Policy represented in versioned, testable, and deployable machine-readable form.

Principal. The human, organization, or authorized actor from which legitimate authority originates.

Privilege amplification. An increase in effective authority as permissions, tools, data, or delegation paths are combined.

Progressive authority. Increasing authority in controlled stages as evidence and operational maturity improve.

Prompt injection. Instructions embedded in untrusted content that attempt to redirect model or agent behavior.

Provenance. Information describing the source, version, custody, and history of data or content.

Rate limit. A restriction on how frequently an actor may perform an operation.

Reauthorization. An explicit decision to restore authority after suspension, incident, or material change.

Reconstruction. Rebuilding the path from principal and task to action and outcome, or tracing backward from an outcome.

Revocation. Withdrawal of authority, credentials, access, or delegation.

Revocation tree. The set of downstream tasks, credentials, and delegations affected by revoking an upstream grant.

RFC – Request for Comments. A publication in the Internet standards and technical documentation series.

Risk-tiered autonomy. Assigning autonomy according to consequence, reversibility, sensitivity, and available controls.

Runtime. The computing environment that executes an agent.

Separation of duties. Dividing initiation, approval, execution, or review among different actors.

Short-lived credential. A credential with a deliberately limited validity period.

SPIFFE – Secure Production Identity Framework for Everyone. Specifications for issuing and verifying workload identities.

SPIFFE ID. A URI-form identifier representing a workload within a SPIFFE trust domain.

SPIRE – SPIFFE Runtime Environment. An open-source implementation of the SPIFFE specifications.

Standing authority. Authority retained independently of a specific task or short execution period.

SVID – SPIFFE Verifiable Identity Document. An identity document used by a workload to prove its SPIFFE identity.

System of record. The authoritative system that establishes and retains a business transaction or state.

Task. The bounded unit of work assigned to an agent.

Task-bound authority. Authority limited to a specific task and its permitted actions, resources, context, and duration.

Task-bound credential. A credential issued specifically for a task and constrained to its authority.

Telemetry. Operational data such as traces, logs, metrics, events, cost, and latency.

Token exchange. Exchanging one security token for another token appropriate to a different actor, audience, resource, or delegation.

Tool. A capability exposed to an agent for retrieving information or performing an operation.

Tool poisoning. Manipulation of tool metadata or returned content to influence later agent behavior.

Transaction authorization. Authorization of a specific consequential action and its parameters.

Transaction ceiling. A maximum permitted amount, volume, or consequence for one or more transactions.

Trust domain. An administrative and cryptographic boundary within which identities are issued and validated.

URI — Uniform Resource Identifier. A standardized identifier for naming a resource or identity.

Volume limit. A restriction on the total number or aggregate value of actions over a period.

Workload. A process, service, container, or other computing unit performing software execution.

Workload API. The SPIFFE interface through which workloads obtain and refresh identity material.

Workload attestation. Verification of a workload using trusted properties of its execution environment.

Workload identity. A verifiable identity assigned to executing software rather than a human user.

Zero trust. An architecture that avoids implicit trust and evaluates access to protected resources explicitly.

APPENDIX B

References

Dates for living specifications and continuously updated guidance are approximate.

1. Booth, Harold; William Fisher; Ryan Galluzzo; and Joshua Roberts. *Accelerating the Adoption of Software and Artificial Intelligence Agent Identity and Authorization.* NIST National Cybersecurity Center of Excellence, initial public draft, February 2026.

   https://csrc.nist.gov/pubs/other/2026/02/05/accelerating-the-adoption-of-software-and-ai-agent/ipd

2. Hardy, Norm. "The Confused Deputy (or Why Capabilities Might Have Been Invented)." *ACM SIGOPS Operating Systems Review*, October 1988.

   https://dl.acm.org/doi/10.1145/54289.871709

3. Hardt, Dick. *The OAuth 2.0 Authorization Framework.* RFC 6749, Internet Engineering Task Force, October 2012.

   https://www.rfc-editor.org/info/rfc6749/

4. Hu, Vincent C.; David Ferraiolo; Rick Kuhn; Adam Schnitzer; Kenneth Sandlin; Robert Miller; and Karen Scarfone. *Guide to Attribute Based Access Control (ABAC) Definition and Considerations.* NIST SP 800-162, originally 2014; updated 2019.

   https://csrc.nist.gov/pubs/sp/800/162/upd2/final

5. Jones, Michael; Anthony Nadalin; Brian Campbell; John Bradley; and Chuck Mortimore. *OAuth 2.0 Token Exchange.* RFC 8693, Internet Engineering Task Force, January 2020.

   https://www.rfc-editor.org/info/rfc8693/

6. Lodderstedt, Torsten; Justin Richer; and Brian Campbell. *OAuth 2.0 Rich Authorization Requests.* RFC 9396, Internet Engineering Task Force, May 2023.

   https://www.rfc-editor.org/info/rfc9396/

7. NIST Center for AI Standards and Innovation. *AI Agent Standards Initiative.* Launched February 2026.

   https://www.nist.gov/artificial-intelligence/ai-agent-standards-initiative

8. Rose, Scott; Oliver Borchert; Stu Mitchell; and Sean Connelly. *Zero Trust Architecture.* NIST SP 800-207, August 2020.

   https://csrc.nist.gov/pubs/sp/800/207/final

9. Saltzer, Jerome H., and Michael D. Schroeder. "The Protection of Information in Computer Systems." *Proceedings of the IEEE*, September 1975.

   https://web.mit.edu/saltzer/www/publications/protection/

10. Tabassi, Elham. *Artificial Intelligence Risk Management Framework (AI RMF 1.0).* NIST AI 100-1, January 2023.

    https://www.nist.gov/publications/artificial-intelligence-risk-management-framework-ai-rmf-10

11. NIST. *Artificial Intelligence Risk Management Framework: Generative Artificial Intelligence Profile.* NIST AI 600-1, July 2024.

    https://www.nist.gov/publications/artificial-intelligence-risk-management-framework-generative-artificial-intelligence

12. SPIFFE Project, Cloud Native Computing Foundation. *The SPIFFE Standard.* Approximately 2017–present; living specification.

https://spiffe.io/docs/latest/spiffe-specs/

13. SPIRE Project, Cloud Native Computing Foundation. *SPIRE Documentation and Architecture.* Approximately 2017–present; living documentation.

https://spiffe.io/docs/latest/spire-about/spire-concepts/

14. Model Context Protocol Contributors. *Model Context Protocol Specification.* Initially released in 2024; continuously revised.

https://modelcontextprotocol.io/specification/2025-11-25

15. Model Context Protocol Contributors. *Model Context Protocol: Authorization.* 2025; evolving specification.

https://modelcontextprotocol.io/specification/2025-11-25/basic/authorization

16. OpenTelemetry Authors, Cloud Native Computing Foundation. *OpenTelemetry Specification.* Approximately 2019–present; living specification.

https://opentelemetry.io/docs/specs/otel/

17. OpenTelemetry Authors, Cloud Native Computing Foundation. *OpenTelemetry Semantic Conventions.* Approximately 2021–present; living specification.

https://opentelemetry.io/docs/specs/semconv/

18. OWASP Cheat Sheet Series Contributors. *AI Agent Security Cheat Sheet.* Approximately 2025–present; living guidance.

https://cheatsheetseries.owasp.org/cheatsheets/AI_Agent_Security_Cheat_Sheet.html

19. OWASP GenAI Security Project. *OWASP Top 10 for LLM Applications 2025.* Published in late 2024 for the 2025 edition.

https://genai.owasp.org/llm-top-10/

20. OWASP GenAI Security Project. *OWASP Top 10 for Agentic Applications for 2026.* December 2025.

https://genai.owasp.org/resource/owasp-top-10-for-agentic-applications-for-2026/

21. OWASP Cheat Sheet Series Contributors. *MCP Security Cheat Sheet.* Approximately 2025–present; living guidance.

https://cheatsheetseries.owasp.org/cheatsheets/MCP_Security_Cheat_Sheet.html

22. OWASP Cheat Sheet Series Contributors. *LLM Prompt Injection Prevention Cheat Sheet.* Approximately 2024–present; living guidance.

https://cheatsheetseries.owasp.org/cheatsheets/LLM_Prompt_Injection_Prevention_Cheat_Sheet.html

APPENDIX

C

Implementation Guides

The reference architecture in Chapter 10 is five regions and three flows. This appendix says what those regions are made of in each major cloud.

It is not a complete landing-zone guide, and it is not permission to treat a platform agent runtime as the authorization system. Each cloud will happily sell you an "agent" that can call tools under a broad runtime role. That is the failure mode the book exists to prevent.

> **NOTICE:**
>
> *These implementation guides will not age well, and may be out of date by the time you read this.*
>
> *Service names as of September 2026. Substitute the current identity, gateway, policy, and telemetry products; do not substitute away the eight workstreams.*

Build the same eight workstreams on every provider:

1. Register the logical agent and bind it to a workload identity.
2. Delegate task-bound authority from a principal without handing over that principal's standing token.
3. Put a policy decision in front of every consequential tool.
4. Enforce that decision at the invocation, not in the prompt.
5. Keep the system of record responsible for the business transaction.
6. Emit a decision trace that can walk forward and backward.
7. Interrupt at task, credential, tool, queue, and agent—not only at the container.

8. Promote authority with an evaluation package, and withdraw it the same way.

Do those eight things and the rest of the platform is interchangeable. Skip them and the brand of the cloud does not matter.

What none of the three clouds will do for you

Cloud IAM answers *which identity may call which cloud API*. It does not answer *whether this purchase, in this amount, for this cost center, under this delegation, may proceed.*

You still have to supply:

- a task and delegation record that outlives the access token;
- a policy that can see tool parameters, not only IAM actions;
- a gateway or PEP that can refuse a call the model has already chosen;
- outcome correlation in the system of record;
- an inventory row that can suspend the running agent.

Treat the cloud identity of the runtime as Chapter 2. Treat the rest of the chain as your architecture.

Keep in mind that these are simply one way of implementing the principles and reference architecture presented in this book. Your organization may have other standards to consider.

Azure

Use Microsoft Entra as the identity plane. Use Azure API Management or an equivalent gateway as the execution boundary. Do not use a single user-assigned managed identity as "the agent."

1. Identity

Create an **Entra Agent ID blueprint** for the logical agent (procurement-agent), not an app registration and not a user account. Microsoft's current guidance is explicit: an app registration is a standard workload, not an agent identity. It has no sponsor and no agent lifecycle.

Create **agent identities** from that blueprint for executions that need their own principal. Require a human sponsor.

Bind the running workload with **workload identity federation** or a **managed identity** on AKS / Container Apps / Functions. The pod or app proves what it is; the blueprint decides which logical agent it may represent.

Separate production and non-production blueprints. A development replica must not be able to obtain the production agent identity.

Retire the blueprint and its child identities when the agent is retired. Disabling a Container App is not identity retirement.

2. Delegation

Authenticate the human principal in Entra ID.

Record the task in your task service (Logic Apps, Durable Functions, or an application database). That record is the delegation.

Issue a **short-lived token** for the agent identity via token exchange / OBO only when the flow is truly on-behalf-of a user and the scopes are the task. For autonomous bounded execution, the agent identity acts as itself under a task-scoped grant, not under the manager's refresh token.

Do not put the manager's access token in session state and let the agent replay it.

3. Policy and enforcement

Write the envelope as policy you can version: approved vendor, category, amount ≤ $2,500, one cost center, no vendor-master writes. Azure Policy governs Azure resources. It does not govern a purchase-order payload. Put that decision in **Microsoft Entra ID Governance / Conditional Access** where it fits, and in an application PDP (Open Policy Agent, Azure API Management policies, or a dedicated decision service) where it does not.

Make **API Management** (or Front Door + APIM) the PEP for `create_purchase_order`. The agent calls APIM. APIM validates the current token, the task identifier, and the decision. The ERP never sees a naked agent call.

Deny by default on every other procurement operation, including vendor-master writes, even if the credential could technically invoke them.

4. Tools and enterprise systems

Register consequential tools as APIM APIs with operation-level policies.

Keep Dynamics 365 / the ERP as system of record. Require an idempotency key on create.

If you use MCP servers, put them behind the same PEP. MCP is a tool transport. It is not an authorization layer.

5. Evidence

Propagate correlation identifiers on every hop: principal, delegation/task, agent identity, instance, policy decision, invocation, purchase-order number.

Send traces to **Azure Monitor / Application Insights** using OpenTelemetry. Keep policy decisions and approvals in an append-friendly store, not only in a debug log.

Retain consequential traces by rule, not by sample rate.

6. Interruption and promotion

Stopping the container is step one. Also: disable the APIM operation, revoke the agent identity's tokens, cancel the Durable Functions / queue orchestration, and issue a compensating cancel against the ERP if the order already exists.

Store the grant (bounded execution ≤ $2,500, evaluation package id, review date) where operations can change it. Entra ID Governance access reviews can expire standing assignments. They will not expire a tool the PEP still allows unless you wire those together.

Amazon Web Services (AWS)

Use IAM for workload identity. Use **Bedrock AgentCore Gateway** (or API Gateway + a policy engine) as the execution boundary. Use **Amazon Verified Permissions / Cedar** as the PDP for tool parameters. Do not give the agent's execution role AmazonBedrockFullAccess and call it an envelope.

1. Identity

Create an IAM role per logical agent where the agents differ in consequence. A shared ECS task role for "all agents on this cluster" is the coarse platform identity Chapter 2 warned against.

Use **IAM Roles for Tasks / IRSA / EKS Pod Identity** so the workload is attested by the platform, not by an access key on disk.

Map the logical agent in your registry to the role ARN it is allowed to assume. The runtime may assume only those roles.

For executions that must be distinct, assume a session with a session name and tags that carry agent_id, instance_id, and task_id. Session tags are not a delegation record. They are binding evidence.

2. Delegation

Authenticate the principal in **IAM Identity Center** or your existing IdP.

Create the task record first.

Use **STS AssumeRole** or a token exchange into a role whose policy is already narrower than the principal. Do not pass the human's Identity Center session token to the agent.

Put task identifier and ceiling in the session context or in a signed delegation record the PDP can read. IAM session duration is not task lifetime.

3. Policy and enforcement

Write Cedar policies that can see tool name and parameters: permit `create_purchase_order` when `context.input.amount <=`

`2500` and vendor is in the approved set; forbid vendor-master writes unconditionally for this agent.

Attach that policy engine to **AgentCore Gateway** or to API Gateway + a Lambda authorizer that calls **Verified Permissions** IsAuthorized. Forbid-wins.

Keep IAM as the coarse gate (this role may invoke this Lambda). Keep Cedar as the transaction gate (this amount, this vendor, this task).

AgentCore Memory and other shared stores need the same treatment. Fine-grained access control on memory is not optional if memory can change later actions.

4. Tools and enterprise systems

Action-group Lambdas and MCP targets sit behind the gateway. They do not sit on the model.

The ERP integration uses an idempotency key and returns the purchase-order identifier.

SCPs in AWS Organizations can block whole classes of action (for example, no agent role in the org may call payments:*). That is a useful floor. It is not the envelope.

5. Evidence

CloudTrail records AWS API calls. It does not record why the agent proposed the call.

Emit OpenTelemetry from the gateway, the PDP, and the tool Lambda. Put task_id, delegation_id, agent_id, and decision_id on every span.

Store the Cedar decision and the ERP document number where an investigator can find them without reading CloudWatch at random.

6. Interruption and promotion

Disable the AgentCore / API Gateway route. Revoke the role session. Drain SQS / Step Functions. Cancel the ERP document if it exists.

Graduated autonomy belongs in the policy store: a forbid on write tools is probation; removing that forbid is the Chapter 9 grant. Changing a prompt is not a promotion.

Google Cloud Platform (GCP)

Use **Agent Identity** on Vertex AI Agent Engine where you run agents there. Use **Workload Identity Federation** everywhere else. Use **IAP for Agents / Agent Gateway** and IAM allow/deny plus a parameter-aware PDP at your APIs. Do not treat a shared runtime service account as the actor.

1. Identity

Prefer the per-agent principal Vertex issues (principal://agents.global.org-.../reasoningEngines/...) over a reusable service account. That principal is closer to Chapter 2's logical-agent binding than `sa-agents@project.iam.gserviceaccount.com`.

Off-platform runtimes use **Workload Identity Federation** from GKE, Cloud Run, or an external issuer. No service-account keys.

Grant IAM roles to that agent principal, not to the compute service account, for the few Google APIs the agent truly needs.

Use **Principal Access Boundary** and IAM deny policies as a hard ceiling: this agent principal can never touch billing export, org IAM, or production deploy buckets, regardless of what a later allow policy says.

2. Delegation

Authenticate the human in Cloud Identity / Workforce Identity Federation / your IdP.

Record the task.

Do not hand the user's Google credential to the agent. If a user-context call is required, use a short-lived, audience-bound token issued for that task.

The Agent Identity auth manager can vault outbound OAuth client credentials. Treat that vault as secrets infrastructure. It is not a delegation record.

3. Policy and enforcement

IAM allow/deny decides whether the agent principal may invoke Cloud Run / a backend.

IAP for Agents on the Agent Gateway can enforce identity and some MCP context. It still will not know that $4,001 is outside the grant.

Put a PDP (IAM Conditions where they suffice, otherwise OPA / a custom decision service) in front of the procurement API and evaluate vendor, amount, cost center, and task state.

The Cloud Run service that talks to the ERP should refuse calls that lack a current allow decision.

4. Tools and enterprise systems

Tools are backend services, MCP servers, or connectors. All of them take identity from the agent principal plus the task.

Keep the ERP as system of record with idempotency.

Memory Bank / sessions on Agent Engine are part of the delegated system Chapter 9 evaluates. Isolate them per principal and task.

5. Evidence

Cloud Audit Logs record Google API use. Add OpenTelemetry on the gateway, PDP, and tool services.

Cloud Trace already speaks OpenTelemetry. Use it for the technical path. Keep policy decisions and purchase-order identifiers in a store you control.

Security Command Center threat detection on Agent Engine is useful. It is not a decision trace.

6. Interruption and promotion

Disable the Agent Engine deployment or strip its IAM bindings. Revoke tokens. Drain Cloud Tasks / Pub/Sub. Compensate in the ERP.

Promotion is an IAM + PDP change tied to an evaluation package, not a new model revision on Vertex.

One sequence, any cloud

Do the work in this order. It is the same order as Chapters 2 through 10.

- Inventory the agents and consequential tools already calling production APIs.
- Put a workload identity on each runtime. Remove long-lived keys.
- Bind logical agent to workload so a shared platform cannot self-assert procurement-agent.
- Stand up a task/delegation record. Stop passing user tokens.
- Put a gateway in front of every consequential tool.
- Write the envelope as policy the gateway can enforce, including amount, vendor, and "no vendor-master writes."
- Turn on idempotency and outcome correlation at the system of record.
- Connect identifiers into one trace.
- Rehearse stop: credential, route, queue, document.
- Only then grant bounded execution, and write the grant into the inventory.

If a cloud service appears that packages several of these steps, use it. Do not skip a step because the package is named Agent.

The test is unchanged from Chapter 11. Create one purchase that should succeed and one that should fail. Walk the first forward from principal to purchase-order number. Walk the second backward from the denial to the policy that issued it. If either walk dies inside the model, the platform is not yet the architecture.

APPENDIX

D

Agentic Enterprise Architectural Chain Review Card

The preceding chapters develop a single architectural model:

Principal → Delegation → Agent Identity → Task → Policy → Tool → Action → Outcome → Evidence

That chain is not a product list and not a sequence diagram. It is the set of relationships that must survive execution if an autonomous actor is going to exercise enterprise authority. Each chapter examined one part of the path. This appendix puts the whole path back in one place so it can be used as a review method.

The test is simple to state and harder to pass. Given one consequential action, can the enterprise follow authority from the principal to the outcome? Given the outcome, can it follow that path in reverse from records it actually keeps? Authentication, policy, and logging can all exist while those connections do not.

The figure on the next page is the model. The pages after it are the method. Use both on a real operation—create a purchase order, issue a refund, change access, merge a pull request, deploy—not on "the agent" as an abstraction.

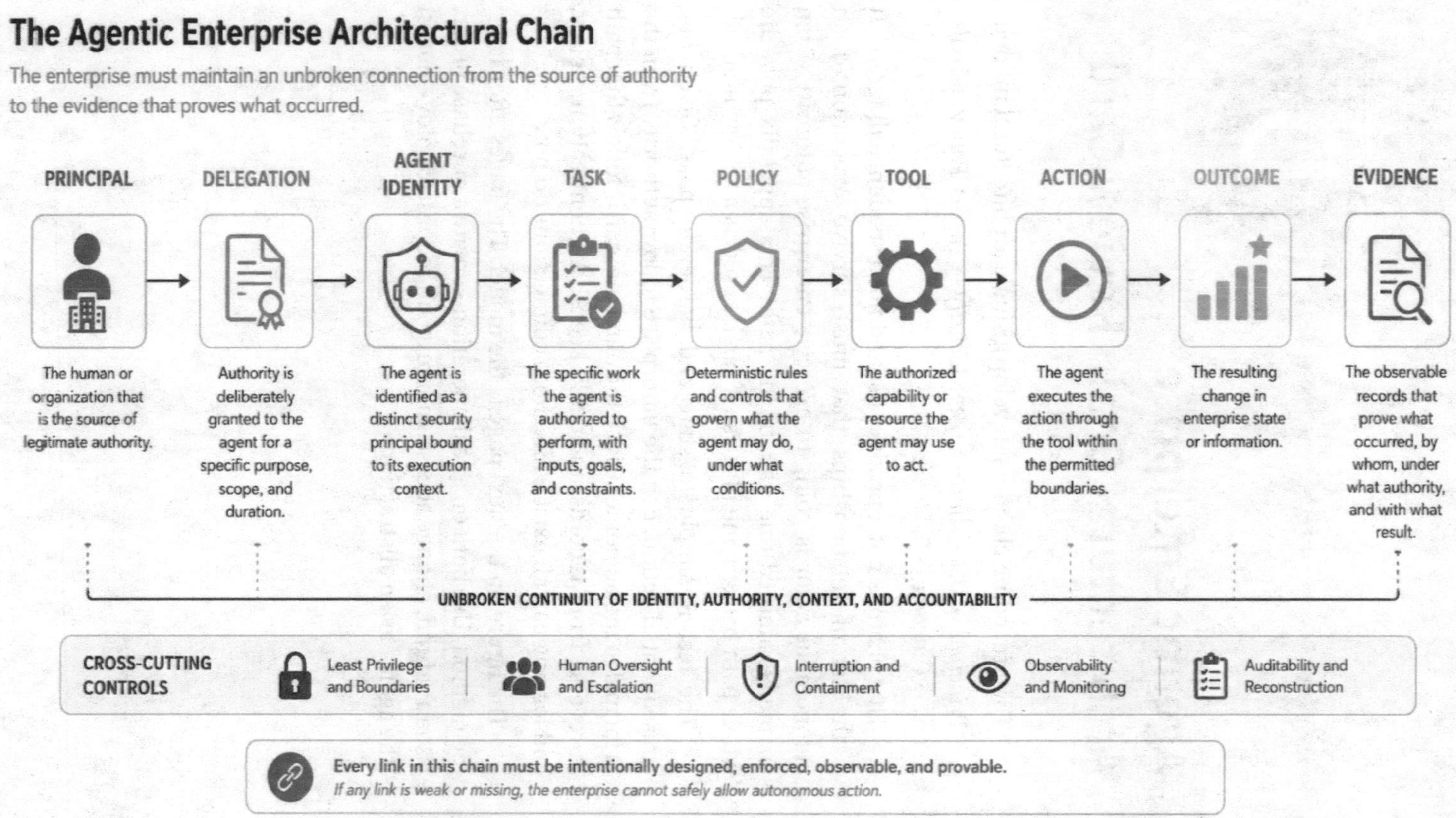
The Agentic Enterprise Architectural Chain
The enterprise must maintain an unbroken connection from the source of authority to the evidence that proves what occurred.
PRINCIPAL
The human or organization that is the source of legitimate authority.
DELEGATION
Authority is deliberately granted to the agent for a specific purpose, scope, and duration.
AGENT IDENTITY
The agent is identified as a distinct security principal bound to its execution context.
TASK
The specific work the agent is authorized to perform, with inputs, goals, and constraints.
POLICY
Deterministic rules and controls that govern what the agent may do, under what conditions.
TOOL
The authorized capability or resource the agent may use to act.
ACTION
The agent executes the action through the tool within the permitted boundaries.
OUTCOME
The resulting change in enterprise state or information.
EVIDENCE
The observable records that prove what occurred, by whom, under what authority, and with what result.
UNBROKEN CONTINUITY OF IDENTITY, AUTHORITY, CONTEXT, AND ACCOUNTABILITY
CROSS-CUTTING CONTROLS
Least Privilege and Boundaries
Human Oversight and Escalation
Interruption and Containment
Observability and Monitoring
Auditability and Reconstruction
Every link in this chain must be intentionally designed, enforced, observable, and provable.
If any link is weak or missing, the enterprise cannot safely allow autonomous action.

1. Name the action.

Pick the operation that changes enterprise state. Not the agent. Not the platform. The action.

2. Walk forward.

- Who authorized the work?
- What authority was delegated, for how long, and with what ceiling?
- Which logical agent—and which execution—received it?
- For what task?
- Which policy decided, and where was that decision enforced?
- Through which tool?
- What action was invoked, with which parameters?
- What outcome did the system of record capture?
- What evidence connects those answers?

3. Walk backward from the outcome.

- What changed?
- Which action produced it?
- Which tool call?
- Which policy decision?
- Which task?
- Which agent and instance?
- Which delegation?
- Which principal?

4. Score the connections, not the boxes.

A platform can have identity, policy, logging, and a tool gateway and still fail this card. The finding is a broken link.

An action with no delegation is standing privilege. A delegation with no bound actor is an identity gap. An action with no policy decision at the enforcement point is self-policing. An outcome with no action identifier means you cannot tell what succeeded. An API log with no principal means you know what happened, not why it was allowed.

5. Decide from the gaps.

Do not fill a hole with a paragraph. Reduce the envelope, require approval, move enforcement to the tool gateway, add correlation, or refuse autonomous execution for that action until the link exists.

The nine names can live in one product or in ten. Least privilege, human escalation, interruption, observability, and reconstruction are not a sixth row of extras after Evidence. They have to be true of every link.

If both traversals can be done from records the architecture actually keeps, the chain holds. If either walk requires a meeting, the chain does not.

www.ingramcontent.com/pod-product-compliance
Lightning Source LLC
LaVergne TN
LVHW030725061126
851857LV00019B/2002

* 9 7 9 8 9 5 2 6 7 7 0 3 6 *